SpringerBriefs in Climate Studies

SpringerBriefs in Climate Studies present concise summaries of cutting-edge research and practical applications. The series focuses on interdisciplinary aspects of Climate Science, including regional climate, climate monitoring and modeling, palaeoclimatology, as well as vulnerability, mitigation and adaptation to climate change. Featuring compact volumes of 50 to 125 pages (approx. 20,000- 70,000 words), the series covers a range of content from professional to academic such as: a timely reports of state-of-the art analytical techniques, literature reviews, in-depth case studies, bridges between new research results, snapshots of hot and/or emerging topics Author Benefits: SpringerBriefs in Climate Studies allow authors to present their ideas and readers to absorb them with minimal time investment. Books in this series will be published as part of Springer's eBook collection, with millions of users worldwide. In addition, Briefs will be available for individual print and electronic purchase. SpringerBriefs books are characterized by fast, global electronic dissemination and standard publishing contracts. Books in the program will benefit from easy-to-use manuscript preparation and formatting guidelines, and expedited production schedules. Both solicited and unsolicited manuscripts are considered for publication in this series. Projects will be submitted to editorial review by editorial advisory boards and/or publishing editors. For a proposal document please contact the Publisher.

Melissa Nursey-Bray • Robert Palmer
Ann Marie Chischilly • Phil Rist • Lun Yin

Old Ways for New Days

Indigenous Survival and Agency in Climate
Changed Times

Melissa Nursey-Bray
Department of Geography, Environment
and Population
University of Adelaide
Adelaide, SA, Australia

Ann Marie Chischilly
Tribal Environmental Professionals
Northern Arizona University
Flagstaff, AZ, USA

Lun Yin
Southwest Forestry University
Kunming, China

Robert Palmer
University of Adelaide
Adelaide, SA, Australia

Phil Rist
Girringun Aboriginal Corporation
North Queensland Land Council Girringun
Aboriginal Corporation
Cardwell, QLD, Australia

This book is an open access publication.

ISSN 2213-784X   ISSN 2213-7858   (electronic)
SpringerBriefs in Climate Studies
ISBN 978-3-030-97825-9   ISBN 978-3-030-97826-6   (eBook)
https://doi.org/10.1007/978-3-030-97826-6

We dedicate this book to all Indigenous peoples across the world who are both impacted by and continue to fight for their land and seas, peoples, cultures, and futures to adapt to the world's greatest challenge, climate change.

We also dedicate this to all our Indigenous friends and relatives who are currently facing or have lost loved ones to COVID, a shock which has made everything that much harder.

Foreword

The subject of climate change is a vast one, and peoples from all across the world are and will be deeply affected by its impacts.

Motivated by a mutual concern about the impacts of climate change specifically on Indigenous peoples, this book is the result of a collaboration between three Indigenous (Ann Marie, Lun and Phil) and two non-Indigenous (Melissa and Rob) colleagues. We have all worked together in various contexts before: Melissa, Rob and Phil co-led a program in Indigenous adaptation for the Social-Economic-Institutional Research Adaptation Network, in Australia, which was funded by the National Adaptation Climate Change Research Network (NCCARF). This work enabled us to gather insights about the different initiatives Indigenous peoples across Australia have been trialling in the climate arena.

Rob and Melissa subsequently travelled to Kunming, China, to visit and work with Lun, who, in his capacity as Professor at the University of Kunming, shared his insights and contacts with us as we sought to understand the experience of ethnic minorities in China in relation to climate change. On another journey, we met with Ann Marie, who introduced us to First Nation colleagues at the Institute of Tribal Environmental Professionals, Arizona, and told us about the impacts of climate change on her people.

We decided to write a book.

To start this process, in 2018, we all met in Hobart, Tasmania, to agree on the narratives that have become the backbone of this book. Along the way, we also collaborated with other colleagues, inviting them to offer additional insights. These are reflected in the series of perspectives that are woven throughout the text. Each perspective represents a 'take' on Indigenous experience of climate change and adaptation. Some are intensely personal, others offer a broader overview, but they all highlight examples of the diversity of Indigenous views, peoples and issues from across the world. Indeed, we do not have examples from every group, and we do not pretend or even try to suggest that this book reflects an 'essentialist' Indigenous view, nor that it represents all Indigenous experience. We are aware and acknowledge we have not been able to highlight as many Asian examples as we would have liked. It is, however, offered as a story that provides a complex series of insights that we hope others can draw upon when facing their own climate change adaptation challenges.

This story shows that adaptation, a term used in Western science to denote a specific response to climate, also embodies, for Indigenous peoples, mechanisms of colonial healing and possibilities for reconciliation. Traditional *and* historical knowledge are shown to play essential roles in establishing adaptation practice. The centrality of building adaptations that incorporate other needs vital to sustaining socio-economic livelihoods within Indigenous communities is acknowledged. Later, our story moves not only to the importance of both navigating competing terminologies but also to the importance of embedding culturally appropriate modes of how to communicate climate change and adaptation in the implementation of adaptation. Finally, we offer reflections on what conditions are needed to obtain institutional or governance 'fit' and enable active agency by Indigenous peoples as they seek to address climate change.

We begin however, with an overview of the idea of climate change and explanations of the key terms used to describe it. This is followed by a summary of how Indigenous peoples across the world have built adaptations to date. We provide a range of examples, from livelihood diversification, adaptation planning, communication strategies, and practices of knowledge revitalisation to name a few.

Our broad analysis is anchored by three contemporary and in-depth case studies from China, the USA and Australia, all of which encapsulate different modes of Indigenous adaptation, agency, and power in practice. Each case study has been chosen to reflect different aspects of Indigenous engagement with adaptation. The case study from China provides insights into how ethnic minorities deploy traditional knowledge systems in agriculture to advance responses to climate change. In our case study of America's First Nations peoples, we introduce the wide scope of geographies, challenges and responses to climate change that are being implemented across the USA. In our investigation of Australia, we show how communications are developed and have fundamental implications for how they can be deployed to build partnerships and contribute to Indigenous-led climate adaptation. The book also provides vivid evocations from other global scholars who have shared their own experiences via a series of climate snapshots.

Ultimately, this book presents Indigenous-driven and led adaptation programs that counter entrenched and dominant discourses about Indigenous vulnerability and resilience. We feel the use of vulnerability creates persistent assertions of victimhood that dilute Indigenous capacity to contribute to policy while romanticised ideas of resilience can be used to erase Indigenous agency and capacity to fully participate in decisions made *about* and *for* them. Despite the depth of the impacts they are facing, we argue that Indigenous peoples are asserting agency and, in deploying their knowledges, are building other ways of being, surviving and knowing in future climate worlds.

Adelaide, SA, Australia	Melissa Nursey-Bray
Adelaide, SA, Australia	Robert Palmer
Flagstaff, AZ, USA	Ann Marie Chischilly
Cardwell, QLD, Australia	Phil Rist
Kunming, China	Lun Yin

Preface I

I am betting a certain number of Indigenous persons have had an awkward experience when they first experienced the concept of climate change articulated by a non-Indigenous politician, educator, scientist, or advocate. For me, I recall gathering from that first articulation some of the following points. Climate change is a phenomenon that is discovered based on the scientific developments of the last few centuries. The phenomena of anthropogenic, or human caused, climate change is accessible primarily to people trained to use particular types of measuring instruments, though positive and negative environmental impacts of climate change will eventually be felt by everyone in the world. Some of the causes of anthropogenic climate change are the burning of fossil fuels, deforestation, industrial urbanisation and transportation, and intensive agriculture.

Such an articulation of climate change comes across as assuming it is the primary intellectual pathway for understanding what is happening to the environment due to the rise in global average temperature. But how can it be the only pathway? For so many Indigenous persons I know from all parts of the world, their most ancient scientific systems emphasised building knowledge about how societies should be organised to be responsive to changing environmental conditions. Changing environmental conditions range from seasonal changes expected to occur each year to longer term trends in environmental conditions across substantive periods of time. Indigenous philosophies that emphasise reciprocity with, and responsibility for relationships with diverse, more than human beings and entities, arise from time tested scientific systems, that owe part of their origin to the need to learn about living well with the dynamics of ecosystems. So, "no", climate change was not "discovered." At least for some Indigenous peoples, what is called climate change today has connections to some of their oldest knowledge institutions.

Across the globe, Indigenous peoples are taking action to address climate change; they are not waiting for invitations to do so from others. Many of the Indigenous persons and their collaborators who authored or are profiled in this book are the leaders taking action now. Their actions are situated in a contemporary moment that is embedded in generations, long histories of lessons learned, and an absolute commitment to the well-being of future generations. This book is about community-based actions to address climate change that centre the lives of diverse people who are committed to collective self-determination, knowledge, creativity, and justice at the grassroots level. Readers will find no narrow or exclusive treatment of climate change here. Readers instead will get a glimpse of the richness of ways to understand and address climate change that have been obscured by so many of the dominant articulations of climate change. Readers will find non-hierarchal and community-wide approaches to what it means to "know" climate change, and the power of reciprocity and responsibility for motivating actions with transformative potential.

School for Environment and Sustainability   Kyle Whyte
University of Michigan
Ann Arbor, MI, USA

Preface II

To embrace readiness for climate change, this volume suggests a radical acceptance of Indigenous knowledge, connections, relationships and practices that have nurtured territories of life for tens of thousands of years. Indigenous peoples hold special and particular knowledge that has long been excluded from research, civic life and decision-making about our global futures. This is to the detriment of our broader societies, as our bio-culturally diverse places call out for multiple ways to care for and steward our precious world.

Indigenous peoples, communities and territories are under increasing physical, emotional and cultural threats to our lands and waters. While the tasks of repair seem overwhelming in their scope, this volume brings a focus to the immediate and shared global concern of climate change. Whether we recognise it, deny it or actively work together to lessen its effects, climate change is now part of our lives and fills us with uncertainty about our futures. Indigenous peoples have long histories of accommodating and valuing change to strengthen our societies and increase health in both people and place. Many of these lessons are relevant and timely to all different populations today.

The editors have provided a culturally safe space for Indigenous peoples to contribute our thoughts and experiences on climate change. We are able to demonstrate the depth of our welcome and commitment to work with others in reciprocal and respectful relationships. When we care for country together, then neither we nor country can be lonely. Changing a climate of fear towards Indigenous knowledge and peoples is an immense challenge, yet the editors and contributors from across many seas are walking together towards the same view. This means we cannot get lost when we are in step with each other.

Centre for Social Impact   Emma Lee
Swinburne University,
Melbourne, Australia
Tebrakunna Country

Photos

Photo 1 Gawler Ranges, South Australia. (Credit: Melissa Nursey-Bray)

Photo 2 The team (from left): Lun Yin, Rob Palmer, Phil Rist, Melissa Nursey-Bray, Ann Marie Chischilly. (Credit: Jack Palmer-Bray)

Acknowledgements

We would like to collectively acknowledge the institutional support we have received that enabled us to work together on this book, including from the University of Adelaide, Australia; Girringun Aboriginal Corporation, Australia; Southwest Forestry University, China and the Institute for Tribal Environmental Professionals, Northern Arizona University, Arizona. We also would like to thank all the Indigenous and non-Indigenous colleagues who agreed to provide perspectives for the book – they add insight and richness and showcase a diversity of perspectives around how to adapt to change. We also thank Dr Masud Kamal for his assistance in editing early drafts and Dr Dianne Rudd for her exhaustive efforts in editing and proofing the later drafts. Finally, for all of us, our work would not be possible without the unceasing work done by Indigenous peoples all over the planet, specifically those with whom we have worked and who have inspired this book – thank you.

Contents

About the Authors[1]

Melissa Nursey-Bray is a researcher with over 20 years of experience working with and for Australian Indigenous peoples, most recently on climate adaptation. Based at the University of Adelaide, Australia, her work considers the role of scale, knowledge and colonisation and how to build Indigenous agency and its incorporation into environmental and climate policy.

Rob Palmer is a communications expert, based in Adelaide, Australia, and wrote a PhD that explores the linkages between climate change adaptation and people from low socio-economic backgrounds. He has real-world experience in advising how to improve science communications about climate change with Australia's Indigenous peoples.

Phil Rist is a Nywaigi man from North Queensland, Australia, and executive officer of the Girringun Aboriginal Corporation. His work includes Indigenous cultural heritage, climate change and governance. He has been an outstanding leader in building cross-cultural partnerships and world-leading co-management programs.

Ann-Marie Chischilly is a renowned environmentalist and past executive director of the Institute for Tribal Environmental Professionals at Northern Arizona University, (NAU) Arizona, USA. She is now Vice President of the Office of Native American Initiatives, NAU. She speaks both nationally and internationally on topics of traditional knowledge, tribes/ Indigenous people and climate change.

Lun Yin is an ethno-ecologist from the Bai ethnic minority community in Kunming, Yunnan Province of China. He is a Professor at the Southwest Forestry University. He is also the director of Centre for Biodiversity and Indigenous Knowledge (CBIK). He is the Founder and Director of the Climate Action SDG Laboratory.

---

[1] This book is a co-authored and cross-cultural collaboration.

Perspectives[2]

Gerald Atampugre is a senior lecturer in the Department of Geography and Regional Planning at University of Cape Coast, Ghana. He undertakes research in the human dimension of climate change, small-holder agriculture and water resource management.

Annette Bardsley works as a researcher in the Department of Geography, Environment and Population at the University of Adelaide, Australia. Her research focusses upon natural resource management, environmental planning and risk management.

Douglas Bardsley is an Associate Professor in the Department of Geography, Environment and Population at the University of Adelaide, Australia. His work explores food security and climate risk and how to adapt to change.

Nicolette Cooley currently serves as one of the co-managers of ITEP's climate change program. She is of the Dine Nation, Arizona (USA), and is of the Towering House Clan, born of the Reed People Clan. Her maternal grandfathers are of the Water that Flows Together Clan and her paternal grandfathers are of the Many goats Clan. Nicolette has a Bachelor and Master of Forestry from the Northern Arizona University and specialises in traditional ecological knowledge.

Karen Cozzetto is one of the co-managers of ITEP's Tribal Climate Change Program and has a PhD in hydrology from the University of Colorado, USA. She was the lead on a Native America Communities and Climate Change Preparedness project at CU, a contributing author on the Indigenous Peoples Chapter of the third National Climate Assessment (NCA3) and a review editor for the Indigenous Peoples Chapter in NCA4.

Haiying Feng works as a Professor in the Qinzhou Development Research Institute at Beibu Gulf University, Qinzhou, Guangxi, China. She undertakes research in the areas of natural resource management, community development and sustainable development strategies.

Lawrence Guodaar is a researcher in the Department of Geography, Environment and Population at the University of Adelaide, Australia. His research focuses on Indigenous adaptation practices, socio-ecological analysis and food security.

---

[2] In this book we provide a number of 'perspectives' which bring vivid evocations of how different First Nations peoples think about and experience climate change. The contributors are listed under "Perspectives".

Niila Inga is the chair of Laevas čearru, a Sámi reindeer herding community in the Kiruna area in Northern Sweden.

Rahwa Kidane is an Assistant Professor in the Institute of Climate and Society at Mekelle University in Ethiopia, and her research focuses on climate change risk perception, climate change impacts, and adaptation and mitigation strategies.

Roselyn Kumar is an adjunct research fellow at the University of Sunshine Coast, Australia. She has extensive fieldwork experience and has published around 40 peer-reviewed articles on climate change adaptation, livelihood sustainability and traditional knowledge.

Emma Lee is an Aboriginal and Torres Strait Research Fellow at the Centre for Social Impact, Swinburne University of Technology, Australia. Her research focusses on Indigenous affairs, land and sea management, policy, and governance of Australian regulatory environments.

Julie Maldonado is a Lecturer in Environmental Studies at the University of California, Santa Barbara (USA). She collaborates with the Institute for Tribal Environmental Professionals to support tribes' climate change adaptation planning. She assists with curriculum development, serves as an Instructor/Lecturer at training sessions and in webinars, and assists with many writing projects.

Md. Masud-All-Kamal is an Assistant Professor in the Department of Sociology at the University of Chittagong, Bangladesh. His research focuses on adaptation governance, social networks and non-governmental organisations.

Karen McNamara is an Associate Professor in the School of Earth and Environmental Sciences at the University of Queensland, Australia. Karen has been undertaking research in climate change adaptation, Indigenous knowledge and gender.

Meg Parsons is a Māori/Lebanese/Pākehā woman and senior lecturer in the School of Environment at the University of Auckland, New Zealand. Her research draws on multiple disciplines. Key areas of research include intersections between multiple social identities, diverse knowledge systems, and values in climate change adaptation planning and actions.

Jasmine Pearson is a researcher in the School of Earth and Environmental Sciences at the University of Queensland, Australia. She is interested in the interactions between mangrove ecosystems and Pacific Island communities.

Brianna Poirier is a Canadian researcher, currently based at the University of Adelaide, Australia. Her research interests include indigenous health, food security and sustainability.

Kristina Sehlin-MacNeil is a Researcher in Várdduo – Centre for Sami Research at Umeå University in Sweden. Her research interests include conflict and power relations between Indigenous peoples and extractive industries and international comparisons of these. She collaborates with Indigenous communities in several countries with a focus on Sweden and Australia.

Victor Squires is a freelance consultant and a former dean of the Faculty of Natural Resources at the University of Adelaide, Australia. He is a well-known environment expert who has conducted many projects in multiple sectors of environment protection, natural resource and biodiversity conservation.

List of Abbreviations

BIA	Bureau of Indian Affairs
CC	Climate Change
COP	Conference of Parties
CTS	Collective Trauma Syndrome
ITC	Inter-Tribal Council
ILO	International Labour Organisation
IK	Indigenous Knowledge
IPCC	International Panel on Climate Change
ITEP	Institute of Tribal Environmental Professionals
LCIPP	Local communities and Indigenous People's Platform
NCA	National Climate Assessment
NCCARF	National Climate Change Adaptation Research Facility
NGO	Non-Governmental Organisation
NBITWC	Norton Bay Inter-Tribal Watershed Council
REDD+	Reducing Emissions from Deforestation and Forest Degradation
TCCP	Tribal Climate Change Program
TK	Traditional Knowledge
TEK	Traditional Ecological Knowledge
UNDP	United Nations Development Program
UNFCC	United Nations Framework Convention on Climate Change
UNU-TKI	United Nations University- Traditional Knowledge Initiative

List of Boxes

List of Figure

List of Perspective

List of Photos

List of Tables

Chapter 1
Introducing Indigenous Peoples and Climate Change

Introduction

Climate change is the challenge of our age. It impacts, in potentially catastrophic ways, Indigenous life worlds, knowledge systems and the environments that they are inextricably connected to (Macchi, 2008; Whyte, 2014; Wildcat, 2013). Indigenous peoples cannot afford to wait for outside help to manage the impacts of climate on their territories and cultures (Alden Wily, 2016; Ramos-Castillo et al., 2017). Further, Indigenous peoples face rising global expectations that their traditional ecological knowledge can be harnessed and used in conjunction with Western science to build stronger and enduring adaptation to change (IPBES, 2019).

Navigating the tension that arises from this attempt to make Indigenous and non-Indigenous interests adhere is a singular challenge, and does not occur in a vacuum, but within, on the one hand, ages old Indigenous stewardship of land and seas, and on the other, a political and historical legacy of dispossession, racism and economic disadvantage, caused by colonisation and globalisation. What role can Indigenous peoples play as they seek both to respond to the impacts of climate change, and yet assert their sovereignty and voice on the international stage as well as within their own territories? It is the exploration of this question, how the world's oldest living cultures are adapting to climate change that is the subject of this book.

The book is underpinned by an exploration of the role that Indigenous knowledge plays in driving various forms of adaptation. There are many definitions of Indigenous knowledge to draw upon, most famously perhaps is Berke's (1999, 8) definition of Indigenous knowledge as Traditional Ecological Knowledge (TEK) which he defines as a "cumulative body of knowledge practice and belief evolving by adaptive processes and handed down through generations by cultural transmission". Indigenous knowledge has also been characterised as a *process*, one that is fluid, dynamic and flexible (Gomez-Baggethun & Reyes-Garcia, 2013), continually evolving, local, culturally specific, and holistic in nature, usually orally transmitted,

M. Nursey-Bray et al., *Old Ways for New Days*, SpringerBriefs in Climate Studies, https://doi.org/10.1007/978-3-030-97826-6_1

and closely related to the survival of a people (Cuerrier et al., 2015; Dudgeon & Berkes, 2003; Pearce et al., 2015; Zimmerman, 2005). Usher (2000) defines TEK as: "all types of knowledge about the environment derived from experience and traditions of a particular group of people". The term "Traditional Climatic Knowledge" has also been suggested (Reedy et al., 2014, 2) and is defined as "an evolving corpus of knowledge that is based on observations and correlations of environmental resource managers concerning climate and climatic changes and the resulting or perceived impact on plants, cultivation and ecosystems."

However, we argue that the deployment of this knowledge is not just about content, nor is it just located in the distant past, but rather that Indigenous peoples are using, maintaining, revitalising, and creating new knowledges, *in the now*, based on traditions of knowledge maintenance going back centuries. We argue that the hallmark of Indigenous adaptation is this skill and an inherent belief in the capacity to continually renew and invigorate culture and knowledge, despite and notwithstanding the impact of colonisation, climate change and globalisation.

Indeed, Indigenous knowledge making did not simply stop with colonisation or other stressors but continues to be gathered and layered. In the context of climate change, we argue that recent historical knowledge, often derived or gathered during periods of colonisation (as much as millennia old knowledge) adds to what is a significant corpus of knowledge about climate impacts, and could be helpful in determining how climate change is understood, communicated and responded to (Nursey-Bray et al., 2020). This sophisticated capacity to adapt knowledge rooted in millennia of experience and apply it to new challenges like climate change, has resulted in a diverse range of adaptation and initiatives which collectively provide insights into how the world may build climate futures. Indigenous knowledge in all its forms is a living thing, and its use and legitimation, however it is spatially constituted in practice and time, can help build decolonised adaptation programs.

In acknowledging all forms of Indigenous knowledge, it is also important to understand that the way language is used to relay that knowledge is important, and further, that when discussing an issue like climate change, it is important to ensure that all parties to the discussion are talking about the same thing. When we discuss 'Indigenous' peoples, what is meant by that? Who are we talking about? When we consider 'adaptation', 'climate change', 'vulnerability' and 'resilience', what is understood by these terms, which are used flexibly and often loosely by different actors. Further, and as will be discussed later, how these terms are deployed and used to advance various dominant knowledge and power agendas are often rejected and countered by Indigenous peoples. Thus, a crucial first step in advancing any conversation about Indigenous adaptation is to clarify what is meant by all these terms.

The Key Terms

'Indigenous'

One of the most important of these of course, is the word "Indigenous". This is because non- Indigenous interpretations of who *is* or *isn't* Indigenous is too often used as a weapon of power to control Indigenous peoples and the flow of resources to them, despite the ways in which they may culturally identify themselves. It is also important because the ebb and flow of resources to Indigenous peoples to address climate change will start with this definition – and thus determine who is or isn't the recipient of any financial support – or inclusion in climate governance by governments and others.

Unsurprisingly, the term 'Indigenous' has been the subject of debate within the United Nations for a long time. Indeed, it is not a term that is used at all in some cases: in Alaska for example the term "Alaska Native"is used while the Constitution of Canada uses the term "Aboriginal". "First Nations" is also a widely used term in Canada, the United States and increasingly in Australia. In Russia, Indigenous peoples are defined according to the size of their population, and groups of less than 50,000 people are legally defined as "Indigenous numerically-small peoples" whereas non-Russian peoples with a population size of over 50,000 are denied Indigenous status. In China, Indigenous peoples are referred to as ethnic minorities and in the Pacific, as "local peoples". In the Asian region, Indigenous Peoples are referred to as tribal peoples, hill tribes, scheduled tribes, janajati, orang asli, orang asal, masyarakat adat, masyarakat hukum adat, adivasis, ethnic minorities or nationalities. In Africa, as shown in the perspective provided by Lawrence Guodaar and Douglas Bardsley below, the term is even more complicated.

Perspective 1.1
Indigeneity in Africa: Implications for climate change adaptation.
Lawrence Guodaar and Douglas K. Bardsley

...being Indigenous to a place is not in itself what makes a people an Indigenous people.
(Barnard, 2006, 1)

The nature of Indigeneity in Africa appears to be more complex than we ever thought. This complexity is not only manifest in the cultural diversity of the varied ethnic and tribal groups, but also how Indigeneity is conceptualised amongst ethnic groups and Indigenous societies in Africa. Migration patterns and their concomitant cultural assimilations of ethnic groups, coupled with the challenges of ancestral home and territorial boundary identification, especially during the pre-colonial period, add to this complexity. As a result, the process of gaining or claiming Indigenous status in many African societies has subtly changed over time. Occupancy of land was one of the approaches by which many African societies established their Indigenous status during the pre-colonial period (Toledo, 2001). During that period, the search for water and pasture for livestock, especially during drought periods generated many of the human migration patterns in Africa. Such demands not only helped many tribal groups adapt to the climatic risks, but currently, traditional mobility is also a means for claiming Indigenous rights to natural resources. Tribal identity is also a conduit to Indigeneity in different jurisdictions, especially in Eastern and Western Africa. For example, in Kenya, Indigenous status can be claimed if one is part of a recognised tribal group, such as the Maasai (Balaton-Chrimes, 2013). Such Indigenous ethnic or tribal groups usually have unique cultures, secret histories and ancestral homes.

Colonisation in Africa however, propelled many Africans to unify and strengthen their cultural identity as a group or groups with a common destiny. From the European colonial era, Africans were generally considered Indigenous in their own right irrespective of their ethnic orientation, while the colonial 'masters' were classified as non-Indigenous. For instance, in Botswana all Africans including the majority and minority groups were considered Indigenous (Pelican & Maruyama, 2015). However, in the post-colonial period, the nature of Indigeneity in Africa, particularly sub-Saharan Africa took on a new dimension. Many ethnic groups have utilised the International Labour Organisation's (ILO) criteria of self-identification as a means of achieving Indigenous status. Thus, different ethnic groups in Africa claim Indigenous status for places where they do not 'originally' belong, in the sense of a pre-colonial heritage with place. Though such groups may claim to be Indigenous in their own right, they are usually not recognised locally, or at least not over all of the claimed areas. Such claims of Indigeneity by migrant settlers could partly be due to their more recent attachment to places and the opportunities they have generated in building livelihood resilience within a local environment.

In view of the anticipated climate change risks in the future, many ethnic groups, and particularly pastoralists in Africa are expected to continue to migrate to different (environmentally favourable) landscapes in an attempt to adapt through improved livelihood resilience. This trend will increase the complexity of claims over land or associated natural resources and how Indigeneity is defined.

There have already been instances where this ambiguity has created competition and conflict between migrant settlers and natives in many regional landscapes in the Sahel (Turner, 2004). The usurpation and control of environmental resources such as land, water, minerals (e.g. gold, oil etc.) and forests by non-Indigenous migrant settlers has the potential to trigger violent tribal conflict (Gleditsch et al., 2007). In fact, it is expected that many vulnerable Indigenous communities in many parts of Africa may continue to be antagonistic towards migrant-settlers through reprisal conflicts. Such conflicts could be exacerbated by the increasing demand from Indigenous people to adapt to climate change through the effective management of the already scarce local resources.

The complexity of the nature of Indigeneity in the African context thus will need to be better understood so as to enable effective decision-making to manage the future challenges of climate change and livelihood disruption. Indigeneity needs to carefully conceptualised within certain defined criteria to generate a broader acceptance within and between African countries. It is imperative that traditional authorities and governments provide these frameworks and criteria to avert conflict between Indigenous and non-Indigenous ethnic groups, to promote peaceful coexistence and sustainable development and build adaptive capacity to respond to climate change (Photo 1.1).

Photo 1.1 Community Gathering in the Afar region Ethiopia. (Credit: Rahwa Kidane)

One of the most commonly used definitions for Indigenous is that of Jose R. Martinez Cobo, the Special Rapporteur of the Sub-Commission on the Prevention of Discrimination and Protection of Minorities, where he notes:

> Indigenous communities, peoples and nations are those which, having a historical continuity with pre-invasion and pre-colonial societies that developed on their territories, consider themselves distinct from other sectors of the societies now prevailing on those territories, or parts of them. They form at present non-dominant sectors of society and are determined to preserve, develop and transmit to future generations their ancestral territories, and their ethnic identity, as the basis of their continued existence as peoples, in accordance with their own cultural patterns, social institutions and legal system (Cobo, 1986, 15).

This definition picks up on the idea of Indigenous groups being *first* peoples, as well as culturally distinct peoples. A long connection with land and sea is another key element - the United Nations Permanent Forum on Indigenous Issues also defines Indigenous peoples as those who have historical continuity, who self-identify at an individual level, have acceptance at a community level, maintain strong links to territories and resources, distinct political, social, linguistic systems, and assert a desire to keep and maintain culture. The International Labour Organisation (ILO), (1989, Article 1, 2) in its Convention concerning Indigenous and Tribal Peoples, defines Indigenous peoples as those:

> People in independent countries who are regarded as Indigenous on account of their descent
> from the populations which inhabited the country, or a geographical region to which the
> country belongs, at the time of conquest or colonisation or the establishment of present
> State boundaries and who, irrespective of their legal status, retain some or all of their own
> social, economic, cultural and political institutions.

Historical continuity is understood to include, occupation of ancestral lands, common ancestry with original occupants, common culture and language, and ongoing residence on that land. Important in this definition is the recognition of colonisation as both a point of *impact* but also the basis on which to define oneself in contrast *to*. The World Health Organisation (WHO, 1999, 1) defines Indigenous peoples as follows: -

> Indigenous populations are communities that live within, or are attached to, geographically
> distinct traditional habitats or ancestral territories, and who identify themselves as being
> part of a distinct cultural group, descended from groups present in the area before modern
> states were created and current borders defined. They generally maintain cultural and social
> identities, and social, economic, cultural and political institutions, separate from the main-
> stream or dominant society or culture.

Other definitions such as that used in Australia, recognise self-identification and acceptance by other members of the relevant cultural group as identifying factors (Department of Aboriginal Affairs, 1981, 2). In New Zealand, the *Maori Affairs Restructuring Act 1989*, the *Rununga Iwi Act 1990* and the *Maori Land Act 1993* define a Maori as a person of the Maori race of New Zealand or a descendant of any such person (Dennis-McCarthy, 2020). In Sweden, the Sami must define himself or herself as Sami and either speak the Sami language as a home language or have a parent or grandparent who spoke the language as a home language.

While these definitions reflect the many different ways by which Indigenous peoples are identified within different nations, the importance of continuity and affiliation with traditional territories, cultural group (self) acceptance, the impacts of forms of colonisation (but not for all groups) and being *first* peoples[1] are shared in common. As a starting point, we suggest that collaborative climate adaptation, wherever in the world it may be, start by recognising how Indigenous peoples identify themselves, not how they are constructed by outsiders.

Western Scientific Terms

Further, the climate change field remains dominated by a Western scientific lexicon (see Table 1.1), and as a result, language continues to play havoc with how climate adaptation can be pursued in cultural contexts.

---

[1] In this book, we acknowledge all these elements, and have decided to use the term 'Indigenous' overall but where we can we also use distinct Indigenous group names.

Table 1.1 Summary of key terms used by the IPCC for climate change

Adaptation	The process of adjustment to actual or expected climate and its effects. In human systems, adaptation seeks to moderate or avoid harm or exploit beneficial opportunities. In some natural systems, human intervention may facilitate adjustment to expected climate and its effects (IPCC, 2014, 1758)
Mitigation	A human intervention to reduce the sources or enhance the sinks of greenhouse gases (IPCC, 2014, 1769)
Climate change	A change of climate which is attributed directly or indirectly to human activity that alters the composition of the global atmosphere and which is in addition to natural climate variability observed over comparable time periods (UN, 1992) A change in the state of the climate that can be identified (e.g., by using statistical tests) by changes in the mean and/or the variability of its properties and that persists for an extended period, typically decades or longer. Climate change may be due to natural internal processes or external forces, or to persistent anthropogenic changes in the composition of the atmosphere or in land use (IPCC, 2014, 1760)
Greenhouse gas	Greenhouse gases are those gaseous constituents of the atmosphere, both natural and anthropogenic, which absorb and emit radiation at specific wavelengths within the spectrum of thermal infrared radiation emitted by the Earth's surface, by the atmosphere itself, and by clouds (IPCC, 2014, 1759)
Adaptive capacity	Ability of systems, institutions, humans and other organisms to adjust to potential damage, to take advantage of opportunities, or to respond to consequences (IPCC, 2014, 1758)
Vulnerability	The degree to which a system is susceptible to, or unable to cope with, adverse effects of climate change, including climate variability and extremes. Vulnerability is a function of the character, magnitude, and rate of climate variation to which a system is exposed, its sensitivity, and its adaptive capacity (IPCC, 2014, 995)
Resilience	The capacity of social, economic, and environmental systems to cope with a hazardous event or trend or disturbance, responding or reorganising in ways that maintain their essential function, identity, and structure, while also maintaining the capacity for adaptation, learning, and transformation (IPCC, 2014, 1772)

This language and these definitions have been used to establish how climate change is talked about world-wide, whether in science, policy or media. Yet, they do not necessarily reflect Indigenous understanding and conceptualisation of the same terms, which is challenging if different parties are trying to work together. These differences in understanding about what key climate terms mean has significant impacts for Indigenous peoples trying to work out how to adapt to climate change and many end up talking at cross purposes with non- Indigenous agencies, governments, policy makers and scientists.

For example, many Indigenous cultures do not have a single phrase such as 'climate change' to reflect their own lifeworld descriptions of the climate, weather and change. Indeed, even the idea that climate change is (a) a scientific process and (b) human induced, is contested by some Indigenous groups who assert that it is God or other ancestors that are causing these changes.

In writing this book with and about many Indigenous groups we have also discovered that the idea of adaptation is *not* constructed by Indigenous peoples as a meteorologically derived and scientific term, but in fact represents a collective cultural adjustment over time to multiple issues and impacts that have affected their knowledge and livelihoods. Time is also constructed differently when discussing the idea of adaptation: for Indigenous peoples, their life worlds construct belonging and identities over a vast timescale, sometimes, as in Australia, over millennia. Climate change and adaptation are thus part of an enduring pattern of survival and existence that has been part of Indigenous lives and world views *always*.

However, Western science, although it does explore climate per se over millennia, constructs *climate change* as the recent acceleration of emissions into the atmosphere and associated warming and other trends. It is understood as a scientific, not a cultural process, and a process that commenced with the industrial scale of burning fossil fuels a mere 250 years ago. Adaptation in a Western context, is adaptation to this recent and accelerated climate change.

Vulnerability and resilience are also understood in quite different ways: the deployment of the Western scientific definitions for them, while setting the benchmark for global understanding of how to construct climate change, have, whether intended or not, been used to entrench dominant constructions of how Indigenous and non-Indigenous peoples react and respond to climate change. As discussed in later chapters, these constructions while familiar tropes in Western management - differ from Indigenous conceptions of what is understood as vulnerability and resilience, thus resulting in climate policy and governance arrangements that do not provide space for Indigenous involvement – and what Indigenous peoples may perceive is appropriate. Many Indigenous peoples for example reject what is a dominant binary construction of them being 'vulnerable' or 'resilient'. Language matters, right from the very beginning.

Adaptation then, which is the core consideration of this book, is not a new concept for Indigenous peoples nor only grounded in climate change contexts. Rather, climate adaptation for Indigenous peoples embodies all the old ways of seeing and doing that have been the basis of their survival for millennia and which now form the foundation of their own, contemporary responses to the challenge of current climate change.

Climate adaptation in this context is not just about the application of selected 'bits' of Indigenous knowledge; nor the naïve incorporation of culturally palatable content: it is about the recognition of Indigenous sovereignty and governance, and a legitimisation of all forms of Indigenous knowledge as a process of survival, an assertion of agency and a vehicle for an ongoing expression of Indigenous culture and healing.

In this context and moving forward, we argue that adaptation, from an Indigenous perspective, begets a completely different approach: for both adaptation that is Indigenous led, and for Western adaptation programs that seek to use Indigenous knowledge. Policy makers need to ensure that Indigenous knowledge is used not just because it is important to the world but do so in ways that also supports Indigenous agency and future worlds (ILO 2017). A world where Indigenous conceptions of adaptation are accepted, and Indigenous voices embedded into governance and planning.

We conclude that adaptation in Indigenous worlds, is not only the application of old ways and knowledge systems to new days but is also an assertion of agency built on patterns of survival that continue to create innovative ways forward.

References

Alden Wily, L. (2016). Customary tenure: Remaking property for the 21st century. In M. Graziadei & L. Smith (Eds.), *Comparative property law: Global perspectives*. Edward Elgar.

Balaton-Chrimes, S. (2013). Indigeneity and Kenya's Nubians: Seeking equality in difference or sameness? *The Journal of Modern African Studies, 51*(2), 331–354.

Barnard, A. (2006). Kalahari revisionism, Vienna and the 'indigenous peoples' debate. *Social Anthropology, 14*(1), 1–16.

Berkes, F. (1999). *Sacred ecology: Traditional ecological knowledge and resource management*. Taylor & Francis.

Cobo, J. R. M. (1986). *Study of the problem of discrimination against indigenous populations, volume IV*. United Nations.

Cuerrier, A., Brunet, N., Gerin-Lajoie, J., Downing, A., & Levesque, E. (2015). The study of Inuit knowledge of climate change in Nunavik, Quebec: A mixed method approach. *Human Ecology, 43*, 379–394.

Dennis-McCarthy, N. (2020). Indigenous customary law and international intellectual property: Ascertaining an effective indigenous definition for misappropriation of traditional knowledge. *Law Review (Wellington), 51*(4), 597–641.

Department of Aboriginal Affairs. (1981). *Report on a review of the administration of the working definition of aboriginal and Torres Strait islanders*. Commonwealth of Australia.

Dudgeon, R., & Berkes, F. (2003). Local understandings of the land: Traditional ecological knowledge and Indigenous knowledge. In H. Selin (Ed.), *Nature across cultures: Views of nature amid the environment in non-Western cultures* (pp. 75–96). Academic Publishers.

Gleditsch, N. P., Nordas, R., & Salehyan, I. (2007). *Climate change and conflict: The migration link. Coping with crisis working paper series*. International Peace Academy.

Gomez-Baggethun, E., & Reyes-Garcia, E. (2013). Reinterpreting change in traditional ecological knowledge. *Human Ecology, 41*, 643–647.

ILO. (1989). *Indigenous and tribal peoples convention, C169*. https://www.ilo.org/dyn/normlex/en/f?p=NORMLEXPUB:12100:0::NO::P12100_ILO_CODE:C169

ILO (International Labour Organisation). (2017). *Indigenous peoples and climate change: From victims to change agents through decent work*. ILO (International Labour Organisation).

IPBES. (2019). In S. Díaz, J. Settele, E. S. Brondízio, H. T. Ngo, M. Guèze, J. Agard, A. Arneth, P. Balvanera, K. A. Brauman, S. H. M. Butchart, K. M. A. Chan, L. A. Garibaldi, K. Ichii, J. Liu, S. M. Subramanian, G. F. Midgley, P. Miloslavich, Z. Molnár, D. Obura, A. Pfaff, S. Polasky, A. Purvis, J. Razzaque, B. Reyers, R. R. Chowdhury, Y. J. Shin, I. J. Visseren-Hamakers, K. J. Willis, & C. N. Zayas (Eds.), *Summary for policymakers of the global assessment report on biodiversity and ecosystem services of the Intergovernmental Science-Policy Platform on Biodiversity and Ecosystem Services*. IPBES secretariat., , 56 pages. https://doi.org/10.5281/zenodo.3553579

IPCC. (2014). Annex II: Glossary in: Climate change 2014: Impacts, adaptation, and vulnerability. In *Contribution of working group II to the fifth assessment report of the intergovernmental panel on climate change* (pp. 1757–1776). Cambridge University Press.

Macchi, M. (2008). *Indigenous and traditional peoples and climate change*. International Union for Conservation of Nature.

Nursey-Bray, M., Palmer, R., Stuart, A., Arbon, V., & Rigney, L.-I. (2020). Scale, colonisation and adapting to climate change: Insights from the Arabana people, South Australia. *Geoforum, 114*, 138–150.

Pearce, T., Ford, J., Cunsolo Willox, A., & Smit, B. (2015). Inuit traditional ecological knowledge (TEK), subsistence hunting and adaptation to change in the Canadian Arctic. *Arctic, 66*(2), 223–245.

Pelican, M., & Maruyama, J. (2015). The indigenous rights movement in Africa: Perspectives from Botswana and Cameroon. *African Study Monographs, 36*(1), 49–74.

Ramos-Castillo, A., Ramos-Castillo, A., Castellanos, E. J., Castellanos, E. J., & Galloway McLean, K. (2017). Indigenous peoples, local communities and climate change mitigation. *Climatic Change, 140*(1), 1.

Reedy, D., Savo, V., & McClatchey, W. (2014). Traditional climatic knowledge: Orchardists perceptions of and adaptation to climate change in the Campania region (Southern Italy). *Plant Biosystems – An International Journal Dealing with all Aspects of Plant Biolog, 148*, 699–712.

Toledo, V. M. (2001). Indigenous peoples and biodiversity. *Encyclopedia of Biodiversity, 3*, 451–463.

Turner, M. D. (2004). Political ecology and the moral dimensions of "resource conflicts": The case of farmer–herder conflicts in the Sahel. *Political Geography, 23*(7), 863–889.

United Nations (UN). (1992). *United Nations framework for climate change convention.* United Nations (UN).

Usher, P. J. (2000). Traditional ecological knowledge in environmental assessment and management. *Arctic, 53*(2), 183–193.

Whyte, K. P. (2014). Indigenous women, climate change impacts, and collective action. *Hypatia, 29*(3), 599–616.

Wildcat, D. R. (2013). Introduction: Climate change and indigenous peoples of the USA. *Climatic Change, 120*(3), 509–515.

World Health Organisation (WHO). (1999). *Indigenous Fact Sheet.* World Health Organisation (WHO).

Zimmerman, E. (2005). Valuing traditional ecological knowledge: Incorporating the experience of indigenous people into global climate change policies. *New York University Environmental Law Journal, 13*(3), 803–846.

Chapter 2
Responding to Climate Change: Why Does It Matter? The Impacts of Climate Change

What *Is* the Impact?

Climate change is occurring across the world and responding to it is increasingly urgent (IPCC, 2021). Indigenous peoples *are* affected by climate change, often disproportionately so, especially given many peoples still rely on and maintain close relationships with their territories. Across the Arctic, the Amazon, the Himalayas and Africa's deserts, Indigenous peoples face huge challenges as they cope with the amplified impacts of vegetation and biodiversity loss, fire, flood, desertification and warming all of which have changed natural and cultural landscapes in unprecedented ways. As diverse as Indigenous peoples, climate impacts occur at multiple levels and scales. In this chapter we explore what these impacts look like, and highlight how differentiated they are at various scales.

Perhaps the most major impact of climate change on Indigenous peoples is the way in which it disrupts traditional knowledge systems which are often based on ecological connections, and tied to place. When those places in turn, change, then the knowledge tied to it faces fracture. Seasonal changes, caused by climate change disrupt Indigenous understandings of how things work, especially when they include changes to the ecological and cultural cues that have historically guided when to undertaking harvesting and/or other cultural traditions. As Kyle (2020) notes

> Climate variations can disrupt the systems of responsibilities [that] community members self-consciously rely on for living lives closely connected to the earth and many living, non-living and spiritual beings, like animal species and sacred spaces, and interconnected collectives like forest and water systems (Whyte, 2020, 600).

For example, in North Queensland, Australia, for the Kuku Yalanji people, when the wattle was in flower, they knew the coral trout would be available for harvest. However, with climate change, the wattles now flower at different times. This is an example of how fundamental climate changes affect the ecological and seasonal cues that are at the foundation of Indigenous knowledge systems, which in this case

© The Author(s) 2022
M. Nursey-Bray et al., *Old Ways for New Days*, SpringerBriefs in Climate Studies, https://doi.org/10.1007/978-3-030-97826-6_2

Photo 2.1 Wattle in flower tells Indigenous groups what times various fish species are running. (Credit: Melissa Nursey-Bray)

have offered guidance about when to hunt and care for the reef and rainforest Country[1] of the Kuku Yalanji for millennia (Photo 2.1).

The Indigenous people in Lachen Valley, North Sikkim in the Himalayas have also observed altered phenology[2] and range shifts in numerous species (Ingty & Bawa, 2012). Some plant species have been found at lower altitudes that were hitherto found at higher altitudes. In other cases, mosquitos have been observed in the higher altitudes due to warming temperatures (Ingty & Bawa, 2012).

Changes to water regimes are also being observed worldwide: for the Apinaje Indigenous people of the Bico do Papagaio region in the Amazon, the drying of the Amazon means that the rain which used to arrive non-stop from January to June, now comes later, and has affected corn and other plantings. The melting of the glaciers in the Himalayas, not only disrupts Indigenous peoples' access to water as snow cover shrinks - but causes floods. In the hills of Zhonzhuang village in

---

[1] "Country" is a term used by Australian Aboriginal and Torres Strait Islanders to denote their traditional land and seas, but also their cultural identity and the rules and laws pertaining to caring for it.

[2] For reference, phenology is refers to the examination of periodic events in biological life cycles, and the ways in which they are influenced by seasonal and interannual variations in climate, as well as habitat factors (such as elevation).

northwest China, climate related water insecurity has caused crop losses, and intensified poverty.

Indigenous agricultural and livelihood practices are being disrupted worldwide. In the Kaduna state of Nigeria climate change is perceived to be responsible for various forms of crop infestations which have reduced both the quality and quantity of the crop yield of local peoples (Ishaya & Abaje, 2008). In Pakistan, centuries of nomadic pastoralism are being challenged as they now have to depend on ad hoc market and cash interactions during seasonal migrations (Xu & Grumbine, 2014). Deforestation and forest fragmentation also affect Indigenous peoples throughout the Amazon, with fires leading to the creation of savannah lands, further disrupting local knowledge and harvesting practices. Nepalese farmers are experiencing changes to indigenous cropping systems, and have started to import new seed varieties, and create protected areas in and around their forest to protect biodiversity (Xu & Grumbine, 2014). In China, the Lisu people, who live in the southwest mountainous areas of the Yunnan province, and are known for their skills in collecting honey from hives on towering cliffs, are finding fewer hives as bee populations decline due to climate change.

Cattle and other forms of herding are being compromised: in Africa, a combination of increased wind, vegetation loss, dune expansion and higher temperatures is affecting cattle and goat farming for the Indigenous peoples within the Kalahari Basin. Throughout Scandinavia, Sami reindeer herders are finding that rain and warmer weather is inhibiting reindeer from finding the lichen they need to survive, causing massive losses of reindeer but also losses to Sami culture and wellbeing. This is an impact being felt by the over 100,000 people in the world who herd reindeer outside of Scandinavia – from Alaska, Canada, China, Greenland, Mongolia, and Russia. For example, Nenets herders from the Russian Arctic have had to delay an annual migration across the Ob River because the ice was not thick enough to cross safely. As Sami leader Olav Mathis Eira notes of the cultural impact of climate changes:

> Many aspects of Sami culture– language, songs, marriage, child-rearing and the treatment of older persons – are intimately linked with reindeer herding…If reindeer herding disappears, it will have a devastating effect on the whole culture of the Sami people (Yeo, 2014, 1).

These stressors mean that climate change also has multiple additional consequences for Indigenous food security (Bryson et al., 2021). Ongoing and previously reliable sources of food are being disrupted: in the Arctic, Indigenous peoples who hunt polar bears, walrus, seals and caribou, herding reindeer, fishing and gathering, are experiencing changes in species viability and availability (Arctic Council, 2013). As Perspective 2.1 highlights, these impacts have multiple ramifications.

Perspective 2.1
Climate Change in Canadian Inuit Communities
Brianna Poirier

Arctic communities across the globe are at the forefront of climate change, with many already experiencing the devastating implications of temperature and sea-level rise, reduced summer sea-ice, glacial melting, coastal erosion and decreased permafrost. Climate change indicators observed by Inuit across Arctic Canada include longer summers, shorter winters and faster thawing of ice; these changes hinder traditionally accurate climate predictions and make travelling on land and ice dangerous (Bonesteel, 2006). These environmental conditions, among others, have severe impacts on Arctic life, culture, food and infrastructure (Andrachuk & Smit, 2012; Furgal & Seguin, 2006). Due to heavy financial and sustenance reliance on hunting and fishing, the Inuit of Inuvialuit have a substantial relationship with both wildlife and weather (Andrachuk & Smit, 2012). Traditional subsistence practices, such as salmon gathering and caribou hunting, are altered when ecosystems are disrupted, which decreases the accessibility to land-based knowledge and language (Markkula et al., 2019).

Tuktoyaktuk, an Inuit community of approximately 900 people, located in the Western Canadian Arctic, within the Inuvialuit Settlement Region of the Northwest Territories, is a good exemplar of these changes. Living on the shores of the Beaufort Sea, residents of Tuktoyaktuk are descendants from the Siglit people, traditionally sea-oriented fur trappers, hunters and herders (Alunik et al., 2003; Usher, 1993). During the early post-contact period, the Inuit who resided around the Beaufort Sea were almost entirely eradicated due to exposure to smallpox and other foreign diseases; this significant loss also resulted in a loss of knowledge particular to the region (Carmack & Macdonald, 2008). Climate change impacts compound these colonial impacts: in Tuktoyaktuk the warming Arctic climate is shifting ecological borders, altering wildlife migration patterns and endangering plant life in tundra climates. Traditional harvesting events of beluga whales and other species revolve around seasonal changes and are reliant on sea ice. Therefore, food security for Tuktoyaktuk community members is at risk due to rising temperatures and subsequent loss of sea ice, in turn causing a decrease in accessibility of country foods (Post et al., 2009), and decreased availability of caribou resulting in significant stress for the people of Tuktoyaktuk. The reproduction cycle of caribou depends on timing of seasonal changes yet warmer temperatures result in earlier vegetative growth - this puts the survival of many herds at risk because caribou migration - which relies on daylight cues - no longer coordinates with vegetation availability (Post et al., 2009; Post & Forschhammer, 2008).

These impacts adversely impact Inuit communities with older populations, high unemployment rates and limited access to retail stores. The spring arrival of whales is another example: impacted by sea ice levels the whales now arrive two and a half weeks earlier than previously experienced. Individuals have also noticed an observed decrease in body mass amongst the whales that arrive. Additionally, community members have cited irregular weather patterns and an inability to predict the weather, which hinders the ability of the community to participate in traditional hunting seasons (Waugh et al., 2018). These changes test the community's capacity to safely and sustainably harvest wildlife, and maintain Indigenous food sovereignty and the cultural values upheld via hunting (The Wainwright Traditional Council, 2011). Further, climate induced coastal erosion affects housing, and infrastructure and housing costs directly influence the ability of families to afford nutrient-dense foods on a consistent basis. This is because all of the infrastructure in Tuktoyaktuk is built on permafrost, making the community vulnerable to coastal erosion and flooding, which are the results of increasing Arctic temperatures, longer summer seasons and increased sea-ice melt. The long-term ramifications of thawing permafrost on the ecosystem are not fully understood, but the subsequent impacts of ocean acidification, coastal erosion and flooding on marine wildlife are of equal concern for community members due to the continued reliance on traditional hunting for food sources (Manson & Solomon, 2007). These climate changes will have profound impacts on the livelihood and wellbeing of Arctic communities.

Yet, Inuit communities have historically persevered in extreme conditions and adapted to significant change via their *Inuit Qaujimajatuqangit* (IQ). IQ collectively refers to traditional ways of knowing, central to all aspects of Inuit livelihood and is especially useful when discussing long-term environmental observations and patterns related to climate change (Bonesteel, 2006).

Utilising IQ in conjunction with scientific knowledge is valuable and helps create a holistic understanding of the relationship between the environment and humans (Markkula et al., 2019). Invaluable information regarding natural ecosystems exists in the experience of Indigenous Peoples in Canada because dependency on the land and its resources for survival develops an intimacy with the environment that many people never experience (Carmack & Macdonald, 2008). The incorporation of IQ in contemporary environmental research will ensure environmental risks and the direction of development are meaningful and accurate for those at the forefront of climate change (Bonesteel, 2006). Homage to historical struggles, resilience and collective desires for healthy, sovereign futures amongst Inuit communities is crucial in developing an understanding of the current impact climate change has on Inuit life (Griffin, 2020).

Such threats to food security as described above, are compounded by the fact that Indigenous populations worldwide also sustain health related impacts from climate change. Numerous case studies of the health impacts of climate change have been documented within the Inuit (Beaumier et al., 2015), the Peruvian Amazon (Hofmeijer et al., 2013), in Canada (Ford, 2012; Furgal & Seguin, 2006; Poutiainen et al., 2013), and Australia (Berry et al., 2010; Bird et al., 2013; Green & Minchin, 2014).

In Canada, for Aboriginal Canadians, existing health challenges will be amplified due to the spread and re-introduction of infectious diseases linked to increased temperatures such as dengue fever, malaria, leishmaniasis, and cholera (Ford, 2012). For example, increased heat will impact practices related to storage and preparation of traditional foods, and thus cause heightened risks of gastroenteritis, food-borne botulism or diseases such as giardia. This is a problem already being experienced by the Batwa people in Uganda (Berrang-Ford et al., 2012; Bryson et al., 2021; Patterson et al., 2017). High rainfall events are also likely to exacerbate exposure to waterborne diseases such as typhoid, or bacillary dysentery. In Canadian, Alaskan, and Amazonian communities, such waterborne diseases are likely to increase diarrheal disease and parasitical infections. In Australia, sea level rise, settlement variability as well as infectious diseases, such as Murray Valley encephalitis and malaria, will impact Indigenous peoples.

Gender inequities are also amplified in climate contexts: for example, many Indigenous women are disproportionately subject to climate related food insecurity and suffer from higher incidences of domestic violence and female displacement because of natural disasters, such as in the Pacific where domestic violence substantively increases after cyclones (Mcleod et al., 2018). Climate droughts in parts of Africa, have led to women and children being forced to walk further to find wood and other supplies. Not only is this time consuming, but also means women have a lot less time to sit down and to pass on their knowledge and teach younger generations. Overall, this gendered terrain of disaster means Indigenous women are more likely to be subject to violence, experience a greater loss of educational opportunities, and experience an increase in their livelihood responsibilities (Bradshaw & Fordham, 2015).

The influence of climate change on gender and health is reinforced by the fact that Indigenous peoples also often live and rely on territories that are particularly physically vulnerable to climate impacts. This is especially challenging for those peoples that have also experienced colonisation (Jones, 2017, 2019; Watts et al., 2018), one legacy of which is ongoing poverty which exacerbates this vulnerability (ILO, 2017; Jones, 2019): Indigenous peoples comprise almost 15% of the world's poor.

Impacts like sea level rise, increased storms, flooding and desertification (caused by water insecurity), are also creating scenarios where Indigenous communities need to consider whether they need to migrate to survive. Sea level rise and coastal erosion in the Pacific and Alaska mean that whole communities might need to be relocated, while in the Amazon, loss of habitat, resulting from climate related wildfire will restrict access to ancestral lands, and motivate migration into urban areas. Apart from the trauma created by enforced moves, such migration will also disrupt and cause knowledge and cultural loss. Further, Indigenous peoples often suffer from additional economic challenges once they have migrated – many for example, do not find employment or face discrimination (ILO, 2017).

This collusion of poverty and migration will also compound the health impacts of climate change on Indigenous peoples. For example, rising heat levels in workplaces will affect workers on lower incomes and cause heat exhaustion, heat stroke and, at times death. These factors will in turn have significant impacts on work productivity, pay and family incomes overall (ILO, 2017). Too often, Indigenous peoples forced to migrate will also end up living in slums, which are, as places, particularly vulnerable to the impacts of climate change, especially natural disasters (ILO, 2017).

Finally, any consideration of the impacts of climate change on Indigenous peoples is not complete without reflecting on its relationship with the impacts of colonisation. In many countries, colonisation resulted in the invasion, dispossession of Indigenous territories, and missionisation, resulting in enforced losses and disruptions to land, seas, knowledge, language and culture. As such, any attempt to try to advance Indigenous climate adaptation has to be situated within a recognition of the contemporary legacy of colonisation which too often mean that attempts to adapt to climate change are confounded by institutional, cultural and legal barriers that inhibit rather than facilitate Indigenous voices and action. In that context, the perspective by Meg Parsons offers a deeply personal reflection from a Maori perspective on the enduring power of this legacy.

Perspective 2.2
Thinking Beyond the Simplistic Accounts of Māori Vulnerability and Climate Change Adaptation in Aotearoa New Zealand
Meg Parsons

Variations of the common Māori saying 'the past is always in front of you' encompasses Māori understandings of time and the importance of knowing one's whakapapa (genealogical connections) and history (Mahuika, 2011). These extend to discussions of the impacts of climate change, and how best to adapt to changing environmental conditions. My hometown of Ōpōtiki (with a population of less than 3000 people of which more than 65% are Māori) currently holds two notable titles: as Aotearoa New Zealand's most socio-economically deprived town and the current homicide capital of the nation. Ōpōtiki is located with the *rohe* (traditional lands and waters) of iwi (tribal group) *Whakatohea* who endured invasion (the British-led colonial military invasion in 1865) followed by the armed occupation and then the unlawful confiscation of their rohe for Pākehā 'settlement' (Walker, 2007). The land on which the town is located was once an interwoven series of landscapes-waterscapes comprised of two rivers (Otara and Waioeka) that meandered and merged together in fertile wetlands, which included lagoons, a diversity of indigenous flora and fauna, as well as Māori communities' kainga (villages) and cultivations.

Today, Ōpōtiki rests unsteadily on coastal floodplains, which are prone to climate related flood and dry out and shake, due to earthquakes. This settler-scape is made up almost entirely of exotic grasses and trees, with more than 90% of the wetlands having been drained, vegetation cleared, and rivers re-engineered to become straightened channels. The township stands enclosed behind stop banks (levees) designed to protect the township and its residents from the regular flood events that occur. This is the landscape that I grew up with: stop banks and floodwaters; frequently muddy and shaky ground; clover, grass, willows, blackberries, and gorse. Sometimes old folks would recall the days when there were giant trees, thousands of birds flying in the sky, and the rivers were filled with tuna (freshwater eels) and inanga (whitebait). Just as sometimes they would also recount to me how 'once upon a time in Ōpōtiki', no one struggled to find work, no one affiliated with a motorcycle gang, no one ducked bullets during gang violence, and no one's mother desperately searched their house for spare change in order to buy her whanau (family) a loaf of bread. Just as with fairy tales, my elders' stories seemed to be in the realm of make-believe rather than my lived reality in Ōpōtiki.

Increasingly scientists, engineers, and government officials warn Ōpōtiki residents and others within the wider Eastern Bay of Plenty region that they are unsafe due to the impacts of climate change and that they must be addressed (Leeder, 2017; Picken, 2019; Whakatane District Council, 2019). Under the new banner of climate change adaptation local government-led actions, both planned and enacted, are increasingly being directed at further hard adaptations, and river improvement and flood defences, once called swamp drainage, are now termed climate change adaptation (Parsons et al., 2019).

Climate change, an apparent 'new' threat to our safety, demands immediate action. Yet, the township is already an unsafe and marginalised space because of the negative consequences of settler colonisation. My own feelings of safety in Ōpōtiki are not, however, primarily related to floodwaters, sea-level rise or any other 'natural' hazard. Instead, my feelings of insecurity, danger, and vulnerability are bound up in the geographies of violence that are a feature of Ōpōtiki, which are never acknowledged by government officials or 'experts' when the topic of hazards and vulnerability are discussed. This violence is everywhere and nowhere (because no one wants to acknowledge it), mostly it occurs in homes, but also extends into the school yards, streets, pubs, community venues, and marae (hapū meeting places). Such geographies of violence are not unique to Ōpōtiki, and are found elsewhere both locally, nationally, and internationally; particularly amongst Indigenous and other minority communities that are subjected to settler colonial policies of dispossession, discrimination, and marginalisation. The violence in Ōpōtiki is fuelled by drug and alcohol abuse and criminal gangs, but is firmly rooted in the endemic poverty (that includes multigenerational unemployment, food insecurity, substandard housing stock, and poor access to health, educational, and social services) that can be traced back to Māori colonial dispossession and marginalisation. Like other places around the world, rates of gender violence (men against women and children) in the town and wider region increase markedly during and after an extreme weather event (Houghton, 2009). I have witnessed first-hand, heart-breaking scenes of Māori women, who fled from floodwaters that engulfed their homes in Edgecumbe in April 2017, subjected to abuse by their male whanau (family) members who were struggling to cope with losses of properties, of economic security, and of their sense of control.

Thus, just as the Otara and Waioweka Rivers, their stop banks, and their oftentimes raging floodwaters are central to my memories of growing up in Ōpōtiki, so too are memories of violence, poverty, and marginalisation in the town. For a long time I just accepted it as the 'normal' and 'natural' way of things, I normalised the stop banks as part of the natural environment just as I normalised the violence and poverty of daily life in Ōpōtiki. It was only when I left did I realise that there was nothing natural about the (settler colonial) way of things.

Accordingly, while dominant scholarship identifies Māori as highly vulnerable to the negative impacts of climate change, it also needs to discuss how this vulnerability is not only a product of biophysical conditions but is more significantly a result of historic and contemporary processes driven by settler colonialism, now also interwoven with globalisation and neoliberalism, and characterised for many Indigenous people by poverty and violence.

We need to consider how Mātauaranga Māori can be made central (rather than marginal) to climate change adaptation policies and plans. *Mātauranga Māori* (the Māori knowledge system) is a relational ontology underpinned by the concept of *whakapapa* (genealogical connections) which binds humans together with non-humans actors (the land/whenua, rivers/awa, seas/moana, biota, climate, and supernatural beings), not only through their daily activities but also through their whakapapa to all things (Harmsworth & Awatere, 2013; Makey & Awatere, 2018; Salmond et al., 2014). Embedded within *Mātauranga Māori*, then, is the principle that what affects the individual (the part) affects the collective (the whole). Further, we need to ensure that poverty-reduction and violence-reduction strategies are similarly at the heart of discussions of sustainable climate change adaptation strategies for Ōpōtiki and elsewhere in Aotearoa. We need to decolonise how we think about climate change adaptation and Indigenous climate change adaptation in particular (Photo 2.2).

Photo 2.2 Rebuilt Flood levee in Edgecumbe, New Zealand. (Credit: Meg Parsons)

Parson's rich insight into the relationship between adaptation and colonisation in Aotearoa is supported by Cameron's (2012, 4) paper on the Inuit where she points out that there is a failure to attend to colonialism as both an historical and contemporary process, and further, in the context of climate change, "the vulnerability and adaptation frameworks risks delimiting the ways in which [northern] Indigenous perspectives, concerns and critiques, can be heard and be effective". Specifically, she talks about the tendency to characterise or place Indigenous peoples as 'local', tied to specific places and territories.

While there is no doubt that cultural identities remain inextricably connected to specific territories, the assertion of the 'local' nature of Indigenous peoples can also be used as a mechanism to "limit the legibility of Indigenous geographies to the realm of the 'traditional' " and in turn "Indigeneity is delimited – only recognised in association with a particular place, hence rendered invisible and without rights to participate in discussions about other places" (Cameron, 2012, 4). In Australia for example, where 75% of Indigenous peoples actually reside in cities, *outside* their traditional country, the focus on their *country* as the dominant conceptual frame within which discussions about Indigenous adaptation are bounded, actually limits the possibilities of being heard in other contexts.

In sum, in this chapter we have described some of the many impacts that will be felt in differentiated ways by Indigenous peoples across the world. While our description is by no means comprehensive, they delineate key areas of focus for action. Impacts on health, gender, food security, hunting, gathering and harvesting practice, poverty, traditional knowledge and mobility are profound and amplified (in some places) by the ongoing impact of colonisation. The challenges ahead are clearly immense and complicated by the fact that they do not exist in isolation. They abut and weave into and within each other to create added complexity when trying to build ways forward.

Nonetheless, Indigenous peoples still possess vast stores of traditional knowledge and connection to their land and seas and an ability to draw on their traditions to innovate in the face of new challenges. Indigenous peoples are crucial agents of change because their livelihood systems, occupations, traditional knowledge and ways of life are essential for combating climate change effectively (Paris Agreement 2015). The exploration of the various ways in which different Indigenous groups assert their agency, reject discourses about their 'lack' and vulnerability and build responses to these changes, is the theme for the rest of this book.

Over the next few chapters, a synthesis of global Indigenous adaptation initiatives is provided, as well as an analysis of the factors that constrain Indigenous people's capacity to build effective adaptation over time. Indigenous peoples still face the challenge of how to build effective dialogue between Western science and Indigenous knowledge systems and ways of doing, as well as working out how to nurture their own community knowledge so that it that remains vital, persistent, and inter-generationally transmitted over time (Krupnik et al., 2018). Many Indigenous peoples see developing these dialogues, and addressing climate change impacts as a challenge to their capacity to maintain their cultural responsibilities to their territories and communities. This idea of responsibility is crucial. In Australia this is called 'caring for country' but across the world, such responsibility is comprised of a collective set of relationships between people and their land and seas, which enable them all to be nourished and cared for. The enactment, and implementation of these responsibilities, is an active and ages old form of adaptation.

In the next chapter, we present some of the ways in which such adaptations, from an Indigenous perspective is being implemented. We argue that these adaptations are rooted, not in recent demands or events, but located within long standing adaptive traditions and knowledge systems, which have enabled Indigenous peoples to respond to a wide range of climate and other environmental changes over historical time.

References

Alunik, I., Kolausok, E., & Morrison, D. (2003). *Across time and tundra*. Seattle.

Andrachuk, M., & Smit, B. (2012). Community-based vulnerability assessment of Tuktoyaktuk, NWT, Canada to environmental and socio-economic changes. *Regional Environmental Change, 12*(4), 867–885. https://doi.org/10.1007/s10113-012-0299-0

Arctic Council. (2013). *Arctic resilience interim report*. Stockholm Environment Institute (SEI) & Stockholm Resilience Centre (SRC).

Beaumier, M. C., Ford, J. D., & Tagalik, S. (2015). The food security of Inuit women in Arviat, Nunavut: The role of socio-economic factors and climate change. *Polar Record, 51*(5), 550–559.

Berrang-Ford, L., Dingle, K., Ford, J. D., Lee, C., Lwasa, S., & Namanya, D. B. (2012). Vulnerability of indigenous health to climate change: A case study of Uganda's Batwa Pygmies. *Social Science & Medicine, 5*(6), 1067–1077.

Berry, H. L., Butler, J. R., Burgess, C. P., King, U. G., Tsey, K., Cadet-James, Y. L., Rigby, C. W., & Raphael, B. (2010). Mind, body, spirit: Co-benefits for mental health from climate change adaptation and caring for country in remote Aboriginal Australian communities. *New South Wales Public Health Bulletin, 21*(6), 139–145.

Bird, D., Govan, J., Murphy, H., Harwood, S., Haynes, K., Carson, D., Russell, S., King, D., Wensing, E., Tsakissiris, N., & Larkin, S. (2013). *Future change in ancient worlds: Indigenous adaptation in northern Australia*. National Climate Change Adaptation Research Facility.

Bonesteel, S. (2006). *Canada's relationship with Inuit: A history of police and program development*. Indian and Northern Affairs Canada.

Bradshaw, S., & Fordham, M. (2015). Chapter 14: Double disaster: Disaster through a gender lens. In *Hazards, risks, and disasters in society* (pp. 233–251). Elsevier Inc.

Bryson, J. M., Patterson, K., Berrang-Ford, L., Lwasa, S., Namanya, D. B., Twesigomwe, S., & Harper, S. L. (2021). Seasonality, climate change, and food security during pregnancy among indigenous and non-indigenous women in rural Uganda: Implications for maternal-infant health. *PLoS One, 16*(3), 0247198.

Cameron, E. S. (2012). Securing Indigenous politics: A critique of the vulnerability and adaptation approach to the human dimensions of climate change in the Canadian Arctic. *Global Environmental Change, 22*(1), 103–114.

Carmack, E., & Macdonald, R. (2008). Water and ice-related phenomena in the coastal region of the Beaufort Sea: Some parallels between native experience and Western science. *Arctic, 61*(3), 265+.

Ford, J. D. (2012). Indigenous health and climate change. *American Journal of Public Health (1971), 102*(7), 1260–1126.

Furgal, C., & Seguin, J. (2006). Climate change, health, and vulnerability in Canadian Northern Aboriginal Communities. *Environmental Health Perspectives, 114*(12), 1964–1970. https://doi.org/10.1289/ehp.8433

Green, D., & Minchin, L. (2014). Living on climate-changed country: Indigenous health, Wellbeing and climate change in remote Australian communities. *EcoHealth, 11*(2), 263–272.

Griffin, P. (2020). Pacing climate Precarity: Food, care and sovereignty in Iñupiaq Alaska. *Medical Anthropology*, 1–15. https://doi.org/10.1080/01459740.2019.1643854

Harmsworth, G. R., & Awatere, S. (2013). Indigenous māori knowledge and perspectives of ecosystems. In J. R. Dymond (Ed.), *Ecosystem services in New Zealand – Conditions and trends* (pp. 274–286). Manaaki Whenua Press.

Hofmeijer, I., Hofmeijer, I., Ford, J. D., Ford, J. D., Berrang-Ford, L., Berrang-Ford, L., & Namanya, D. (2013). Community vulnerability to the health effects of climate change among indigenous populations in the Peruvian Amazon: A case study from Panaillo and Nuevo Progreso. *Mitigation and Adaptation Strategies for Global Change, 18*(7), 957–978.

Houghton, P. (2009). *People of the great ocean: Aspects of human biology of the early Pacific.* Cambridge University Press.

ILO (International Labour Organisation). (2017). *Indigenous peoples and climate change: From victims to change agents through decent work.* ILO (International Labour Organisation).

Ingty, T., & Bawa, K. S. (2012). Climate change and indigenous peoples. In M. L. Arrawatia & S. Tambe (Eds.), *Climate change in Sikkim patterns, impacts and initiatives* (pp. 275–290). Information and Public Relations Department, Government of Sikkim.

IPCC (2021). Summary for policy makers 2021: The physical science basis. Contribution of working group 1 to the sixth assessment report of the intergovernmental panel on climate change (Masson-Delmotte, V., Zhai, A., Pirani, S., Connor, C., Pean, S., Berger, N., Caud, Y., Chen, L., Goldfarb, M., Gomis, M., Hunag, K., Leitzell, E., Lonnoy, J., Matthews, T., Maycock, T., Waterfield, O., Yelekci, R., Yu and Zhou, Cambridge University Press.

Ishaya, S., & Abaje, I. B. (2008). Indigenous people's perception on climate change and adaptation strategies in Jemma LGA of Kaduna State. *Journal of Geography and Regional Planning, 1*(8), 138–143.

Jones, A. D. (2017). Food insecurity and mental health status: A global analysis of 149 countries. *American Journal of Preventive Medicine, 53*(2), 264–273.

Jones, R. (2019). Climate change and indigenous health promotion. *Global Health Promotion, 26*(3), 73–81.

Krupnik, I., Rubis, J. T., & Nakashima, D. (2018). Indigenous knowledge for climate change assessment and adaptation: Epilogue. In *Indigenous knowledge for climate change assessment and adaptation* (pp. 280–290). Cambridge University Press.

Leeder, D. (2017). Doug Leeder, Bay of Plenty Regional Council Chair, speech given at media conference at the release of the Rangitaiki River Scheme Review Report. Speech presented at the *Findings of the independent review into Rangitaiki River Scheme, Whakatane.* https://www.boprc.govt.nz/latest-news/media-releases/media-releases-2017/october-2017/findings-of-the-independent-review-into-rangitaiki-river-scheme/. Accessed 8 Oct 2020.

Mahuika, N. (2011). 'Closing the Gaps': From postcolonialism to Kaupapa Māori and beyond. *New Zealand Journal of History, 45*(1), 15–32.

Makey, L., & Awatere, S. (2018). He mahere pāhekoheko mō kaipara moana–integrated ecosystem-based management for kaipara harbour, aotearoa New Zealand. *Society & Natural Resources, 31*(12), 1400–1418.

Manson, G., & Solomon, S. (2007). Past and future forcing of Beaufort Sea coastal change. *Atmosphere-Ocean, 45*, 107–122.

Markkula, I., Turunen, M., & Rasmus, S. (2019). A review of climate change impacts on the ecosystem services in the Saami Homeland in Finland. *Science of the Total Environment, 692,* 1070–1085. https://doi.org/10.1016/j.scitotenv.2019.07.272

Mcleod, E., Arora-Jonsson, S., Masuda, Y. J., Bruton-Adams, M., Emaurois, C. O., Gorong, B., & Whitford, L. (2018). Raising the voices of Pacific Island women to inform climate adaptation policies. *Marine Policy, 93,* 178–185.

Parsons, M., Nalau, J., Fisher, K., & Brown, C. (2019). Disrupting path dependency: Making room for Indigenous knowledge in river management. *Global Environmental Change, 56,* 95–113.

Patterson, K., Berrang-Ford, L., Lwasa, S., Namanya, D. B., Ford, J., Twebaze, F., et al. (2017). Seasonal variation of food security among the Batwa of Kanungu, Uganda. *Public Health Nutrition, 20*(1), 1–11.

Picken, D. (2019, September 2). *Big water: Planning for climate change, extreme weather and rising seas in the Bay of Plenty.* Bay of Plenty Times. https://www.nzherald.co.nz/nz-herald-in-depth/news/article.cfm?c_id=1504170&objectid=12201672. Accessed 4 Sept 2019.

Post, E., & Forschhammer, M. (2008). Climate change reduces reproductive success of an Arctic herbivore through trophic mismatch. *Philosophical Transactions of the Royal Society B, 363,* 2369–2375.

Post, E., Pedersen, C., Wilmers, C., & Forschhammer, M. (2009). Warming, plant phenology and the spatial dimension of trophic mismatch for large herbivores. *Philosophical Transactions of the Royal Society B, 275,* 2005–2013.

Poutiainen, C., Berrang-Ford, L., Ford, J., & Heymann, J. (2013). Civil society organizations and adaptation to the health effects of climate change in Canada. *Public Health London, 127*(5), 403–409.

Salmond, A., Tadaki, M., & Gregory, T. (2014). Enacting new freshwater geographies: Te Awaroa and the transformative imagination. *New Zealand Geographer, 70*(1), 47–55.

The Wainwright Traditional Council. (2011). *Climate change, food, and 'sharing' among the Iñupiat of wainwright, Alaska.* Cornell University. [Online]. Available: http://www2.dnr.cornell.edu/kassam/publications/Climate_Change_Food_and_Sharing_in_ainwright_Alaska.pdf

UNFCCC. (2015). Paris Agreement to the United Nations Framework Convention on Climate Change, Dec. 12, 2015, T.I.A.S. No. 16–1104.

Usher, P. (1993). Northern development, impact assessment, and social change. In N. Dyck & J. B. Waldram (Eds.), *Anthropology public policy and native peoples in Canada* (pp. 98–130). McGill-Queen's University Press.

Walker, R. (2007). *Opotiki-Mai-Tawhiti: Capital of Whakatohea.* Penguin Books.

Watts, N., Amann, M., Ayeb-Karlsson, S., Belesova, K., Bouley, T., & Boykoff, M. (2018). The Lancet Countdown on health and climate change: From 25 years of inaction to a global transformation for public health. *Lancet, 391,* 581–630.

Waugh, D., Pearce, T., Ostertag, S., Pokiak, V., Collings, P., & Loseto, L. (2018). Inuvialuit traditional ecological knowledge of beluga whale (Delphinapterus leucas) under changing climatic conditions in Tuktoyaktuk, NT. *Arctic Science, 4*(3), 242–258. https://doi.org/10.1139/as-2017-0034

Whakatane District Council. (2019). *Climate change.* Whakatane District Council. https://www.whakatane.govt.nz/climate-change. Accessed 4 Sept 2019.

Whyte, K. (2020). Too late for indigenous climate justice: Ecological and relational tipping points. *Wiley Interdisciplinary Reviews. Climate Change, 11*(1).

Xu, J., & Grumbine, R. E. (2014). Integrating local hybrid knowledge and state support for climate change adaptation in the Asian Highlands. (Report) (Author abstract). *Climatic Change, 124*(1–2), 93–104.

Yeo, S. (2014). *Five ways climate change harms indigenous people, in Climate Home News.* Online web site, https://www.climatechangenews.com/2014/07/28/five-ways-climate-change-harms-indigenous-people/. Accessed 10 May 2021.

Chapter 3
Indigenous Adaptation – Not Passive Victims

Introduction

A Mind Set, a Process: Asserting Voice and Action

Indigenous peoples are not passive victims of climate change and indeed have been adapting in multiple ways to climate change for millennia. As Makondo and Thomas (2018, 89) note: "to Indigenous communities there is nothing new about climate change. These communities have lived with and adapted to it for centuries". In this chapter we present the ways in which Indigenous peoples have responded to the current challenge of climate change, and we explore the dynamic ways in which Indigenous knowledge is deployed to build adaptive responses. We also present other forms of adaptation which include the development of climate adaptation strategies, knowledge revitalisation and maintenance programs, science partnerships and governance initiatives. At its core, this range of adaptations, show that Indigenous adaptations are more than the application of adaptation *content*, they reflect a mindset, an adaptation *process*, one that relies on ages old practice, and that creates spaces for the assertion of contemporary Indigenous voices and agency.

Indigenous Knowledge and Adaptation

The question of how Indigenous peoples can contribute to climate mitigation and adaptation efforts is a core deliberation for Indigenous peoples and policy makers across the world. This is because the perceived capacity of Indigenous knowledge to combat climate change, has created mounting pressure on Indigenous peoples to share their knowledge for the common good (Hill et al. 2020). This is an ambition clearly identified in the Intergovernmental Science-Policy Platform on Biodiversity and Ecosystem Services (IPBES), which has established a task force on Indigenous

© The Author(s) 2022
M. Nursey-Bray et al., *Old Ways for New Days*, SpringerBriefs in Climate
Studies, https://doi.org/10.1007/978-3-030-97826-6_3

and local knowledge systems to assist in the identification of ways to work with Indigenous and local knowledge to address biodiversity loss and climate change. As it states:

> Indigenous peoples and local communities possess detailed knowledge on biodiversity and ecosystem trends. This knowledge is formed through their direct dependence on their local ecosystems, and observations and interpretations of change generated and passed down over many generations, and yet adapted and enriched over time. Indigenous peoples and local communities from around the world often live in remote areas, interacting with nature and managing resources that contribute to society at large.... They are often better placed than scientists to provide detailed information on local biodiversity and environmental change and are important contributors to the governance of biodiversity from local to global levels (IPBES, 2019, 1).

This international recognition of Indigenous knowledge is important as it offers an opportunity for Indigenous peoples across the world to participate in international agencies and programs such as the IPBES. However, while Indigenous knowledge may be important to others, and plays a role in the global quest to redress biodiversity loss, the maintenance of that knowledge also needs active support, with structures put in place to protect Indigenous cultures, territories and knowledges in their own right. Systems need to be put in place to ensure Indigenous knowledge is not appropriated and further damaged (where globalisation and colonisation have already had enormous impacts), and that provision is made for Indigenous peoples to assert their voice within institutions and climate change and adaptation governance regimes and programs.

International Governance

Finding voice is integral to achieving successful adaptation, thus it is unsurprising that given climate change impacts are serious and far reaching, Indigenous peoples have remained staunch in their demands to be part of the decision making about how to respond to the challenges they face. Indigenous peoples are now part of a wide range of international governance agencies and structures concerned with making decisions on a global scale about climate change.

Indigenous approaches to mitigate and adapt to climate change have been articulated for decades with early examples including the Quito Declaration on Climate Change in 2000, the Marrakech Statement on Climate Change from Indigenous Peoples and Local Communities Caucus of the Seventh Session of the Conference of the Parties, and the United Nations Framework Convention on Climate Change in 2001. These early declarations asserted the need both to recognise Indigenous knowledge and contributions to climate change management, but also the rights to participate in global climate governance.

In 2008 the International Indigenous People's Forum on Climate Change was established to be a caucus for Indigenous peoples participating in the United Nations Framework for Climate Change Convention (UNFCCC), providing an important avenue for Indigenous voices on climate to be heard. In 2009, Indigenous representatives from all over the world met in Anchorage, Alaska to discuss the impacts of climate change on their peoples, resulting in what is known as the *Anchorage Declaration of the Indigenous People's Global Summit on Climate Change*. As noted below (UNFCC, 2009, 1), this declaration highlights the urgency and scale of climate change for them:

> We express our solidarity as Indigenous Peoples living in areas that are the most vulnerable to the impacts and root causes of climate change. We reaffirm the unbreakable and sacred connection between land, air, water, oceans, forests, sea ice, plants, animals and our human communities as the material and spiritual basis for our existence. We are deeply alarmed by the accelerating climate devastation brought about by unsustainable development. We are experiencing profound and disproportionate adverse impacts on our cultures, human and environmental health, human rights, well-being, traditional livelihoods, food systems and food sovereignty, local infrastructure, economic viability, and our very survival as Indigenous Peoples. Mother Earth is no longer in a period of climate change, but in climate crisis. We therefore insist on an immediate end to the destruction and desecration of the elements of life.

In 2013, the International Indigenous Peoples' Forum on Climate Change (IIFPCC, 2014, ix) made statements at the 38th sessions of the Subsidiary Bodies to the UNFCCC in Bonn that:

> we insist that non-carbon benefits and non-market approaches should be supported in all aspects of the process and should be interconnected with the UNFCCC REDD+ safeguards as agreed to by the Parties in Cancun.

This was followed in 2014, with a statement from International Indigenous Peoples Forum on Climate Change (2014, 2) that:

> 1. Parties shall ensure that Paris agreement respects, protects and fulfils the human rights of Indigenous peoples including their rights to lands, territories and resources as enshrined in the UN Declaration on the Rights of Indigenous peoples (UNDRIP) and 2. The subsistence livelihoods of Indigenous peoples are key to biodiversity conservation and enhancement, to ensure food security for millions of people, and contribute to climate change adaptation and mitigation. Therefore, Indigenous peoples are encouraged to share the good practices of Indigenous agricultural systems in SBSTA discussion on agriculture in its 43rd session.

In 2016, the Paris Climate Change Agreement recognised that Indigenous people must be the part of the solution to climate change. Importantly, in 2018, the Local Communities and Indigenous Peoples Platform (LCIPP) was established to build the capacity for engagement of Indigenous peoples and local communities; and to design climate change policies and actions in a manner that respects and promotes the rights and interests of Indigenous peoples and local communities.

In 2019, the Yogyakarta Declaration, a result of a gathering of Indigenous peoples from Cambodia, Myanmar, Philippines, Taiwan, Thailand, Timor-Leste, and Vietnam, made clear statements about the need to recognise Indigenous knowledge systems, food and cultural security and human rights in the context of climate change.

The emergence of these Indigenous voices at the international scale into climate negotiations about mitigation and adaptation is important (Reisinger et al., 2014), and creates key institutional mandates that faciliate Indigenous voices in international governance arrangements and decision making. An ongoing challenge however, is in working out how such high-level activity can be tailored to address local situations where Indigenous peoples can and are adapting to change on the ground.

Local Place and Knowledge-Based Adaptation

In these contexts, Indigenous peoples have been active in applying their knowledge to address the climate impacts they are experiencing in their land and seas. As described earlier in the book, much Indigenous knowledge is based on seasonal observation, and climate change is responsible for a range of seasonal disruptions, which in turn disrupt traditional livelihood practice. However, at the same time, Indigenous knowledge and processes are now being used to help adapt to these changes.

The Inuit for example apply their knowledge about sea ice conditions, weather and wildlife to better understand climate change and how to respond to it (Gearheard et al., 2010; Laidler et al., 2009; Nichols et al., 2004). In Mongolia, local people's knowledge and observations about climate has been combined with climate science to generate adaptation plans (Marin, 2010). In the Arctic, Indigenous peoples have been documenting their climate histories and knowledge about climate change (e.g. Laidler, 2006; Leduc, 2007; Nickels et al., 2006; Riedlinger & Berkes, 2001; Weatherhead et al., 2010): using that knowledge as a wellspring and spring board for adaptation (Ford et al., 2006, 2008, 2016; Pearce et al., 2015; Reid et al., 2014; Roué, 2018). Indigenous groups in the Andes such as the Charzzani people in the Apolobamba region in Bolivia, have knowledge that helps them to read the weather and understand climate patterns, and thereby adapt to management practices (De la Riva et al. (2013), while in Mexico, the Zoque people have built local climate calendars (Sánchez-Cortés & Chavero, 2018).

For the Yolngu people in Blue Mud Bay, Australia, traditional knowledge is deployed to (re) read the signs in nature and so ensure the keeping of ancestral knowledge (Barber, 2018). At the other end of the world, the Sami have designated 318 noun stems to identify various types of snow which they use as a unique knowledge base to adapt to climate related snow changes (Mathiesen et al., 2018).

Such agency has also been used to generate a wide range of livelihood diversification for many Indigenous peoples. In Nigeria for example, adaptation innovations address current issues by drawing on past practice. Some examples of adaptation include soil and water management, the use of improved/different varieties of crops, planting cover crops like melon, instilling zero tillage, regular weeding, early planting, mulching, use of organic manure and adapting which seeds are selected to plant (Nzeadibe et al., 2012). The Fulani people of Northern Ghana, focus on Indigenous adaptation strategies which create socially just and sustainable ways for herdsmen to maintain their cattle. Strategies include mobility-based strategies, diversification of sources of feed, labour division for meeting differentiated needs of cattle, and stress management in cattle (Napogbong et al., 2021). In Latin America a vast array of adaptation strategies for livelihood diversification are being implemented by multiple local Indigenous peoples. In all these examples, and as Table 3.1 shows, the use and adaptation of the application of local knowledge plays an important role.

Table 3.1 Indigenous livelihood diversification strategies to adaptation in Latin America

Local knowledge-based adaptation strategies	Sources
Andean farmers adapt to unpredictable and changing weather conditions by dispersing cultivated plots across the community's territory instead of a single continuous plot	Boillat et al. (2013)
Farmers use selected tree species at specific locations to adapt to changing flood patterns in the Amazon delta	Vogt et al. (2016)
Small scale fishers-farmers in Amazon floodplains adaptation to climate change use a wide range of local ecological knowledge including planting crops later than before, using small boats, and moving the cattle to upland earlier, to reduce their vulnerability to climate variability	Oviedo et al. (2016)
Using phenomenological markers (presence of cloud, soil moisture), Ch'ol farmers in Chiapas, Mexico, adapt to the changing rainfall patterns by planting local varieties of crop instead of hybrid ones, adjusting plantation times and cultivating multiple crops at a time	Briones (2018)
Indigenous women in Taupi, Nicaragua, use strategies such as traditional recovery therapy for the plant, reinstatement of traditional collective practices such as the *pana-pana* and change in diet, in order to adjust to climate viability and change.	Fenly (2018)
In response to climate change, smallholder farmers in Central America use a variety of adaptation practices including change of crop varieties, agroforestry and intensification. Coffee and basic grain farmers plant more trees on farms	Harvey et al. (2018)
Aymara communities in the Altiplano are introducing new crops, changing varieties of potatoes and using new cropping techniques as adaptation strategies	Kronik and Verner (2010)

In the Cameroons, where 85% of the local Indigenous peoples rely on agriculture for their survival, context specific agro-pastoral adaptations are also being implemented which include crop diversification and mixed farming (Azibo & Kimengsi, 2015).

For some groups, such as in Iran, more structural solutions are being explored: the Qashquai people have implemented long-term adaptive solutions such as building cement block houses and water storage ponds, as well as making changes to their migration routes and location so they can to cope with the adverse consequences of climate change (Saboohi et al., 2019). In the Chittagong Hill tracts of Bangladesh, the Garo people have similarly implemented more structural climate adaptations (Rahman & Alam, 2016), which collectively help reduce the risk of food security and build community resilience to climate change. Others focus on physical changes to landforms: Cree hunters in Canada for instance have installed adaptations to coastal uplift which have included the construction of mud dykes and the cutting of tuuhiikaan, (which are corridors in the coastal forests), to retain and enhance desirable conditions for goose hunting (Sayles & Mulrennan, 2010).

Overall whether it is the implementation of adaptations to crops, water, soil or technology, the active use of Indigenous knowledge to adapt to climate impacts in these examples underpins all modes of adaptation. These adaptations demonstrate the active and contemporary engagement of that knowledge, in the here and now, to new situations, and highlight the fluidity and adaptability, not just of the knowledge itself, but its capacity to regenerate and renew under new circumstances. In this way, the deployment of knowledge as a process as much as its content reveals the living nature of Indigenous cultures and their ability to navigate and survive enormous change.

The advantage of being 'on the spot' and being the repositories of local knowledge and experience, also means that Indigenous peoples have been able to put in context the effect of climate change impacts at the local scale. Their observation of impacts at the local level often correlate with and complement the wider scientific predictions, and enables the generation of more strategically localised responses by both Indigenous peoples and policy makers. Indigenous knowledge in this context becomes a form of place based cultural capital that can help withstand shock and becomes the strength of ongoing cultural life, a source from which adaptation programs can spring. Indigenous knowledge is thus used to inform a continuum of adaptation that integrates management and spirit. Maori led adaptation initiatives in the South Island Hapu, Kati Huirapa demonstrate this fluidity and adaptability. Here the Maori have navigated states of transition in their lands and seas, to integrate management, spirit and ecosystems via continual states of transition and renegotiation (Carter, 2019).

Another example is that of management of traditions around the bowhead whale hunted by the local Inupoaq people in the Arctic which "integrates all elements of Arctic life: that is the seas the land, the animals and humanity. The bowhead remains central to Inupoaq life" (Sakakibara, 2018, 265). In this case, the Inupoaq draw on their cultural strength and ages old cultural practice to build resilience to climate change and in so doing they re-affirm cultural identity, and in strengthening those bonds, use the strength of their past to adjust to the future (Sakakibara, 2018, 267). Case studies from Africa (also see perspective 3.1 below) also show how different peoples use location specific knowledge to combat the effects of climate change (Makondo & Thomas, 2018). As such, Indigenous knowledge is "neither singular nor universal but rather a voluminous, diverse and highlight localised source of wisdom" (Makondo & Thomas, 2018, 73) where "spirituality comprising traditional practices has far reaching benefits in resilience building in many cultures" (Makondo & Thomas, 2018, 85).

Perspective 3.1
Climate Change Adaptation in Rural Ghana: Indigenous People as Change Agents?
Gerald Atampugre

Indigenous peoples are key change agents in the fight against climate change in Ghana. An in-depth understanding of the land as well as their ingrained resilience, drawn from their culture and traditional knowledge, means that climate adaptation can be built and succeed at local levels. If Indigenous cultures flourish, climate change adaptation will succeed at the local level.

Ghana's Indigenous communities and peoples are diverse, span the entire country and are unique in their own right. However, many tend to live in marginal rural environments that are most prone to climate change impacts (e.g. communities along the coast, drylands, etc.) and rely on renewable natural resources for food, cultural practices, and socio-economic activities. They experience higher levels of (non-) climate-related risks compared to other groups and face huge barriers in terms of access to basic livelihood assets (usually they are considered the poorest of the poor, with high gender inequality). Thus, rural Indigenous people in Ghana, like in many other parts of the world, are regarded as victims of climate and other forms of change.

However, Ghanaian Indigenous cultures and traditional knowledge is resilient and adaptive and has already survived several decades of anthropogenic-induced climate and environmental change. Collectively held knowledge of the sea, land and sky, means rural Indigenous Peoples are exceptional observers and interpreters of changes in climate and their surrounding environment. This accumulated knowledge offers invaluable insights, complementing scientific data with context-specific details that provide a crucial foundation for adaptation and the building of resilience in indigenous households and communities.

Indigenous people in Ghana are not passive victims and relative to other groups, they possess unique characteristics that make them important agents for sustainable climate adaptation in Ghana. One such attribute is the fact that they share a complex religious-cultural relationship with the natural resources they depend on for their livelihood endeavours. Indigenous Knowledge Systems protect the natural environment on which they derive their productive assets and incomes (e.g. fruits, mushrooms, bush meat, fish, roots, medicine and construction materials, etc.). For instance, for the Bonos people in forest areas (an Indigenous tribe in Bono East Region of Ghana), physical and spiritual survival is dependent on the depth of their connection with nature i.e. the forests/mangroves and water bodies that are the abodes of their gods and ancestors.

Consequently, there are particular days in the week no one is allowed to enter the forest/mangrove (for timber, firewood, farming purposes or any purpose) or go to the stream/river for water or fishing. These practices support the replenishment of natural resources and biodiversity, leading to improved ecological services. Further, Indigenous ecological knowledge underpins environmental stewardship strategies in rural communities in Ghana. In these rural settings, customary law is recognised and enforced. Over the years, these Indigenous rules and regulations have helped, to some extent, to ameliorate local development and means that communities can withstand, recover from, and adapt to risks.

The unique nature of traditional knowledge and cultural approaches among indigenous people in Ghana thus situates them as important change agents today. To reduce maladaptation and adaptation deficits among Indigenous People in Ghana, the current planned neoliberal top-down adaptation approaches must shift to acknowledge autonomous Indigenous adaptation strategies as building blocks to, and not 'other' strategies. A solid understanding and policy incorporation of Indigenous adaptation practices in Ghana presents an opportunity for a re-evaluation of existing knowledge to enhance context-specific climate change adaptation.

Knowledges Working Together

The development of knowledge partnerships is another way in which Indigenous knowledge can provide place-based information that science alone cannot achieve. The development of partnerships also provides an opportunity for Indigenous peoples to become equitably involved in adaptation projects, from their inception through to delivery, thus maximising Indigenous voices and agency, and often enabling employment. The establishment of partnerships can also redress inequalities in power relations, where Indigenous peoples may be historically underrepresented (Wheeler et al., 2020).

Indigenous peoples are already leading such knowledge collaborations with scientists and policy makers, with examples in the Arctic (Pennesi et al., 2012), the Asian Highlands (Xu & Grumbine, 2014) and Africa (Grey et al., 2020) to name a few. A specific example in the Kw Zulu Natal (Basdew et al., 2017) in Africa combines Indigenous and Western scientific knowledge and seasonal (Indigenous farming) indicators in the Swayimane Mshwathi Munupyu region, Kwa Zulu in South Africa to ensure farming resilience. In another African example, farmers in Malawi have been using Indigenous knowledge for centuries to help them predict and understand weather patterns and hence make decisions about farming. However, given that climate change is destabilising the certainty of that knowledge, they are now also working with scientists to help them predict future weather events and build their resilience via the integration of both knowledge systems to adapt to climate change (Kalanda-Joshua et al., 2011). In Vietnam, the Yao people use their Indigenous knowledge about native crop varieties and animal breeds, weather forecasting, and the timing and location of cultivation practices, both on its own and with science to build resilience to climate impacts (Son et al., 2019).

Indigenous groups across Australia are also working on knowledge partnerships with scientists and governments to re-introduce fire burning as a cultural adaptation strategy. Fire burning was traditionally used to help establish and maintain hunting grounds, to crack rocks, to create tools and weapons, to maintain trade routes, keep travel corridors open, clear water ways and to ensure seed germination. For example, in the State of Victoria, the local Gunditjmara people collaborate with staff from the Forest Fire Management Section of the Department of Environment, Land, Water and Planning, and the Victorian Country Fire Authority, to undertake cultural fire management as part of an overall fire management plan for the Budj Bim National Park (Commonwealth of Australia, 2020). In so doing they assist with fuel reduction. In Australia this is important: as increases in the frequency and intensity of bush fires is predicted, the reduction of fuel load is needed to reduce the risk and likelihood of catastrophic fires while enabling Indigenous peoples to look after and sustain their traditional country (Commonwealth of Australia, 2020).

Such collaborations are growing in number and strength across the world, and while these are just a few examples they reflect some of the diversity and possibilities for partnerships that can offer two-way benefits and learning if resourced and implemented appropriately.

Plans and Strategies

The development of specific adaptation plans and policies, both Indigenous led and undertaken in partnership with others is another adaptation option increasingly trialled by Indigenous groups. In these plans, Indigenous peoples not only offer their knowledge, but also assert/embed their own forms of cultural governance and agency, in turn assuring they are culturally resonant and have currency with their own people.

In the United States, multiple First Nations peoples have created and are implementing place-based adaptation plans, and Chap. 4 provides a detailed description of many of these initiatives. In Australia, both Aboriginal and Torres Strait Islander peoples have prepared adaptation strategies. The Arabana people, from south central Australia for example, have developed an adaptation strategy that identifies adaptation as something that is wider than climate. The Arabana people seek not only to address climate change but position adaptation as a form of healing from colonisation. Adaptation is presented as an opportunity to address past injustices, and to build economic livelihoods as part of the adaptation process. The Arabana also offered a range of insights for policy makers and planners in relation to what may be specific factors worth inclusion in Indigenous adaptation plans (Box 3.1 below).

Box 3.1 Criteria for Inclusion in Indigenous Adaptation Plans

1. Acknowledge that adaptation in Indigenous contexts cannot be extricated from the history of colonisation.
2. When conducting assessments and adaptation, find ways of integrating both the vulnerability of people and the vulnerability of country.
3. Embed justice and equity in all adaptation options and policy.
4. Recognise that adaptation needs to incorporate the lived reality of Indigenous peoples, that they live in urban as well as remote places, adaptation policy needs to integrate place and scale.
5. Document and identify other drivers or agents of change such as mining, and their impact on and implications for adaptation policy and planning.
6. Integrate knowledge sets. Integration is not just about integrating science and knowledge but also re-vitalising knowledge and acknowledging history and memory as a form of knowledge.
7. Co-design adaptation in ways that will assist building community capacity.
8. Develop communications about climate change not only at the local scale; Indigenous people can engage with international dimensions.
9. Ensure flexible governance arrangements that facilitate coexistence between formal and informal and traditional and Western laws and mores thus assisting effective development and implementation of adaptation.
10. Build a trans-disciplinary approach into research.

In the Torres Strait, (which is a group of islands off the top of Australia), sea level rise and migration are active considerations. To address these issues an adaptation strategy has been developed which embeds traditional knowledge as the cornerstone for adaptation, with a commitment to work with government agencies and others to develop evidence-based adaptations for people, the environment and settlements. Relocation is stated as being a last resort (TSRA, 2014).

These are just a few examples of many Indigenous led adaptation plans, but they illustrate how Indigenous leadership can galvanise action across communities, even when, as a result of colonisation, they may be geographically dispersed as a people.

Training and Education Programs

Indigenous peoples have also created innovative and dynamic forums by which they communicate climate change both within their own communities but also to educate policy makers, scientists and others. These forums include the development of digital platforms, training tools, web sites and various education and awareness programs. In Canada for example (see Box 3.2 below), there are a range of communication resources available to assist Indigenous peoples to adapt to climate change.

Box 3.2 Communication Resources to Assist Indigenous Adaptation in Canada

Indigenous Climate Hub: This web site seeks to provide a platform to Indigenous peoples across Canada to share their climate change experiences and stories. It is led by Indigenous peoples and provides resources and tools for others seeking to build adaptation programs

Climate Change Adaptation Planning Toolkit For Indigenous Communities - Centre for Indigenous Environmental Resources
This is another Canadian resource that provides the following: (i) Indigenous Climate Change Adaptation Guidance document, (ii) Climate Change Adaptation Planning Guidebooks for Indigenous Communities, (iii) Indigenous Languages Glossary Workbook and (iv) Two Indigenous Language Glossaries, all designed to assist with adaptation planning https://yourcier.org/

Retooling for Climate Change: a site providing a compendium of resources that include adaptation tools and resources for Indigenous Nations, local governments and others to prepare for the impacts of climate change https://retooling.ca/

Plan2Adapt: This is a tool for generating maps, plots, and data describing projected future climate conditions for regions throughout British Columbia and assists Indigenous peoples to assess climate change in their region based on a standard set of climate model projections. https://toolkit.bc.ca/tool/plan2adapt/

In Bali, Indonesia, climate field schools have enabled the delivery of adaptation actions for local farmers, where agriculture is the main form of livelihood and is threatened by climate change (Biskupska & Salamanca, 2020). Several other adaptation projects also showcase diverse modes of communication as Indigenous peoples try to engage their own people in understanding and responding to climate change. One of these, the Indigenous Youth Climate Art Contest, launched by Fraser Council, in British Columbia, asked Indigenous youth to submit art pieces that showed one or more of the following elements:-

- Valuing and protecting the land and water
- Connecting with and honouring Indigenous knowledge
- Celebrating community and personal resilience
- Understanding impacts to self, community and nations
- Exploring key challenges
- Self-determination
- Innovative responses to climate change.

This competition successfully engaged young Indigenous individuals to create images of climate change but also to learn about it. The artwork from this contest was then showcased in Canada's 2020 national assessment report on climate change titled *Canada in a Changing Climate*, and also helped educate the wider public about Indigenous perspective on and how they were affected by climate change. Another initiative called SEED, is an Australian Indigenous Youth Network that represents an alliance of Aboriginal and Torres Strait Island young people who seek climate justice alongside the Australian Youth Climate Coalition. They run campaigns and disseminate their activities via a dynamic web site.

Indigenous Climate Action also provides a comprehensive set of resources, podcasts, activities and resources that support Indigenous peoples the world over to combat climate change. It does so via four pathways - gatherings, resources and tools, amplifying voices and Indigenous sovereignty. In their website, they have an 'amplifying voices' section which provides a comprehensive array of reports and documents which assert and document Indigenous perspectives on key issues. One report for example, is of a youth assessment, which outlines youth views and aspirations relating to climate change, giving voice to Indigenous youth in ways not usually possible in wider forums.

Influencers and Drivers Affecting Indigenous Adaptation

As we have shown, Indigenous peoples across the world, are being active change agents and are building climate adaptations in diverse ways. Yet, climate adaptation is not a homogenous exercise, and in attempting to implement adaptation of various kinds, there are also a range of challenges and barriers, often external non climate related stressors, that may prevent their effective implementation. As we showed in Chap. 2, any discussion of climate impacts cannot be separated from how they in turn interweave with the legacy of colonisation. But for some Indigenous peoples, one of the most significant factors affecting their capacity to adapt is also the way in which Indigenous ways of being and knowing have been impacted by forms of globalisation which drive a variety of external non-climate related stressors that affect Indigenous peoples. Given its impacts include the appropriation/extraction of land and culture and often enduring socio-economic disadvantage, globalisation is a process that has been constructed by many as a different type of colonisation. As such, globalisation as much as colonisation remains an active agent in the navigation of climate adaptation in Indigenous worlds. The complex way in which colonisation, globalisation and climate change interact is illustrated in the perspective below, where forms of (colonially derived) extractive violence hinder the capacity for Sámi reindeer herders to adapt over time.

Perspective 3.2
Sámi Reindeer Herding, Extractivism and Climate Change
Kristina Sehlin MacNeil and Niila Inga

Climate change is often spoken of as something that *will* happen, for reindeer herding Sámi communities it is something that *is* happening and that already *has been* happening for many years. Indigenous communities, living in close connection with their lands and environments, are at the frontier when the effects of climate change are felt. Today, climate change affects Sámi reindeer herding in direct and destructive ways, so much so that some communities find it difficult to see a future for their livelihoods.

Sámi people are the Indigenous people of Sweden, Norway, Finland and the Kola Peninsula in Russia. Sápmi is the name of the Sámi homeland which covers parts of these countries. The Sámi were recognised as an Indigenous people in Sweden in 1977 and as a people in the Swedish constitution in 2011 (Hansen & Olsen, 2006; Reimerson, 2015). The Sámidiggi, Sami Parliament of Sweden, was inaugurated in 1993 and in 2000 the Sámi languages were protected through national language legislation (Lantto & Mörkenstam, 2015; Pietikäinen et al., 2010). An estimated 20,000–40,000 Sámi people live in Sweden; however, these figures are not conclusive as Sweden does not identify ethnic groups in official statistics (Axelsson, 2010).

There are 51 reindeer herding Sámi communities in the Swedish part of Sápmi and an estimated ten per cent of Sámi people in Sweden work with reindeer husbandry (Samer, 2021). Since the Swedish implementation of reindeer herding legislation in the late 1800s, Sámi customary and property rights are tied to the reindeer and each community is allocated specific grazing lands. Furthermore, Swedish law governs how many reindeer each community is allowed, based on calculations of land capacity and how many animals that particular areas can sustain (Allard & Brännström, 2021).

For the Sámi the reindeer holds immense importance. Every part of the animal is put to use and the reindeer is central to Sámi history, society and culture, evidenced in part by its presence in Sámi creation stories and ancient as well as contemporary expressions of art (Gaski, 2008). There is an annual cycle in reindeer herding, following the seasons. This differs both locally and regionally, as different communities experience different climate zones, in addition there are mountain Sámi communities and forest Sámi communities that practice reindeer herding in slightly different ways. However, a common goal for all reindeer herding Sámi communities is to ensure that the reindeer have access to enough pasture lands and the best possible grazing within the various communities' customarily held areas. The reindeer wander freely and are thus dependent on natural grazing lands, they feed on certain types of lichen commonly found in the northern parts of Sweden. The reindeer is a migratory animal and it is therefore, important for the reindeer herds to be able to move from pasture land to pasture land depending on seasons.

Climate change is increasingly affecting herding practice. For example, reindeer herders have been forced to move their reindeer herds by lorry in order to ensure the safety of their animals as they need to cross service roads or train tracks - or lakes that will no longer freeze over due to the milder climate caused by climate change.

In Laevas čearru, one of the 51 reindeer herding Sámi communities in Sweden and located in the Kiruna area of Northern Sweden, some reindeer herders have taken to feeding their reindeer in enclosed paddocks. This is due to the severe effects of extractive industries and climate change, which has made it increasingly difficult for the reindeer to find enough food in winter. Others are forced to buy fodder and distribute to the reindeer on their natural pasture lands. Apart from the inconvenience and costliness of feeding the reindeer this way, it also causes problems for the animals as their digestion is sensitive and not adapted to food types other than lichen. There is also an increased risk for reindeer to contract diseases as they need to be gathered up in larger groups for support feeding to be effective.

Herders now seek to understand what kind of adaptations will be necessary for Sámi reindeer herding to survive the current climate and that of the future (REXSAC, 2021).

However, the Swedish part of Sápmi is attractive to extractive industries: there are large forest areas, mineral rich ground, and several large rivers, often referred to as the last wilderness in Europe. Forestry, mining, hydropower, and tourism are all industries that want access to Sámi customary lands and the places where reindeer roam and forage. Thus, where climate change has made reindeer herding increasingly difficult to manage, the expansion of extractive industries is an additional concern that often threatens to become the final straw that will break the livelihood of reindeer herding as it means that land needed for herding is appropriated for extractive use: "Every square meter of land there is becomes incredibly important if we are to be able to adapt to the current climate changes", says Niila Inga, chair of Laevas čearru. These conflicting interests between reindeer herding Sámi communities and extractive industries and entrepreneurs are often difficult to resolve and the Sámi perspectives are often trivialised or ignored in consultation processes (Sehlin MacNeil, 2015, 2017).

Extractive violence – that is, "a type of direct violence against nature and/or people and animals that is caused by extractivism and that primarily affects peoples closely connected to land" (Sehlin MacNeil, 2017: 23) – underpinned by structural and cultural violence, happens in situations of asymmetric power relations or conflicts. These conflicts, where Indigenous peoples are involved, have their roots in colonial structures and attitudes. Where profit driven extractive ideologies are dominant, Indigenous perspectives on connections between land and people, are commonly trivialised or ignored (Sehlin MacNeil, 2017). However, Indigenous peoples' knowing, being and doing in relation to the lands that they care for and manage *has* been given more space in public debate and Indigenous protests about the destruction of their lands are numerous and global. Some are loud and visible, some silent and strategic, some pass quickly, others remain for years on end. There are serious efforts made, also within the frameworks of research, to understand Indigenous cultures from non-colonial perspectives, where living in reciprocal relationships with the land, not trying to own, destroy or master it, is described as affluent and intelligent economies (Broome, 2010; Gammage, 2011). Through recognising extractive violence against land as violence against people, proponents of extractive ideologies can hopefully begin to understand the severe impacts such activity has on our shared planet (Photo 3.1).

Photo 3.1 Grazing Paus moving East. (Credit: Niila Inga)

What Sehlin-MacNeil and Inga show in this perspective is how other drivers such as mining for iron ore, intensive use of rivers, clear felling of forests, dams and the creation of national parks also restrict the spaces available for adaptation. This experience is not unique – in Bolivia the livelihoods of Indigenous farmers in the Bolivian highlands are increasingly threatened due to the effect of multiple stressors that accumulate with climate change. Such stressors include land scarcity, uncertainties in agricultural and labour markets, institutional marginalisation, in addition to climate-related stressors such as water shortages, rising temperatures and increased climate variability. The cumulative result of the combination of these factors is that farmers have lost their incomes, experience food insecurity and suffer from reduced levels of natural, human, financial, physical and social capital (McDowell & Hess, 2012). In Southeast Asia, where there are over 150 million Indigenous Peoples, rapid development combined with climate change is having an immense impact on livelihoods and wellbeing. As outlined in another perspective below, external stressors add to the complicated challenge for the local Afar Indigenous peoples of Ethiopia when managing their resources for climate change.

Perspective 3.3
Climate Change Adaptation by the Afar Indigenous People of Ethiopia
Rahwa Kidane

The Afar indigenous people live in the Dankil desert, in the Afar region of Ethiopia. The region is mainly a pastoral area, and it is among the hottest and driest places on earth with temperatures sometimes rising above 50 °C (Cavalazzi et al., 2019). Over the past decade, climate-induced extreme droughts and heatwaves have hit the Afar region more frequently than ever (Afar National Regional State, 2010; Tilahun et al., 2017). In response, the Afar Indigenous people are traditionally adapting to these extreme events mainly through pastoralism (e.g. selling of livestock, storing of dried meat) and mobility to nearby rangelands, guided by their traditional ecological knowledge (TEK) systems (Balehegn et al., 2019; Eriksen & Marin, 2014). In doing so, however, the Afar people are facing developmental processes and policy challenges that threaten their Indigenous climate change adaptation responses.

One of the developmental challenges faced by the Afar Indigenous people relates to the loss of their drought grazing lands to large-scale irrigation farming such as cotton and sugar cane plantations (Fratkin, 2014; Gebeye, 2016; Schmidt & Pearson, 2016). This development measure has led to a massive loss of livestock in Afar, which in turn increases the vulnerability of the Indigenous people to climate change and undermines their capacity to respond to its impacts (Magnan et al., 2016). In addition, although pastoralism is the preferred adaptation measure and one that has a great cultural significance for the Afar, Ethiopia's National Adaptation Plan of Action (NAPA) and the Afar Regional State Adaptation policies promote investments in crop cultivation and other non-pastoral adaptation strategies (Afar National Regional State, 2010; NMA, 2007). Most government officials in Ethiopia regard pastoralism as a "backward" culture, and one that needs to be transformed into "modern" agriculture, in order to "civilize" pastoralists (Müller-Mahn et al., 2010). Such adaptation policies have forced the Afar people to shift out of pastoralism and engage in marginal livelihood activities and unsustainable coping measures (e.g. trade and charcoal production) (Eriksen & Marin, 2014; Magnan et al., 2016).

The example of the Afar clearly reveals that beyond external factors such as climate change, development measures and adaptation policies have reinforced vulnerability by disrupting Indigenous peoples' adaptation strategies and ways of living. This implies that for successful adaptation, there is a need to recognise the context within which vulnerability occurs, including the ones created by non-climatic drivers. The root causes of vulnerability need to be addressed in conjunction with climate change to support and strengthen Indigenous adaptation measures. This case also reveals the presence of a clear disconnect between the government's climate-change adaptation policy measures and that of Indigenous peoples' interests and cultural values. This underscores the need to align the existing top-down adaptation policy measures with Indigenous adaptation priorities and the need to include them in the policy formulation and implementation processes (Photo 3.2).

Photo 3.2 Afar Tribe Man with his camel. (Credit: Rahwa Kidane)

Another issue, related to both the impacts of colonisation and the ongoing prevalence of external globalising pressures such as mining and agriculture on Indigenous territories, is that of tenure. Both colonisation and extractive development pressures, have caused intense pressure on Indigenous lands. In the context of places like Australia, New Zealand, the USA and Canada, Indigenous peoples have also been forcibly relocated and dispossessed of their land and seas, which have been appropriated for various forms of land use, including residential settlement. In other cases, significant Indigenous territories (Yellowstone in the US is an early example) have been taken over and declared National Parks. The historical relationships between governments and its peoples have caused ongoing confusion or injustice in relation to securing tenure for Indigenous peoples. In the Philippines, the resilience of the Higaonon communities in Initao and Naawan, is inhibited by the absence of land entitlements, also exposing them to encroachment by logging and other companies on their ancestral lands (Peña et al., 2017). This insecurity of tenure has wide ramifications as it creates an obstacle to the development of effective adaptation, a fact recognised formally by the International Panel for Climate Change (IPCC) in 2019 as follows:

> Insecure land tenure affects the ability of people, communities and organisations to make changes to land that can advance adaptation and mitigation (*medium confidence*). Limited recognition of customary access to land and ownership of land can result in increased vulnerability and decreased adaptive capacity (*medium confidence*)

The IPCC (2019) then follows up by arguing that

> Land policies (including recognition of customary tenure, community mapping, redistribution, decentralisation, co-management, regulation of rental markets) can provide both security and flexibility response to climate change (*medium confidence*).

Building tenure security then is an important precursor to building strong Indigenous adaptation. It can also play a key role in reducing emissions, as Indigenous lands are recognised globally as an important carbon sink. It is also recognised that many Indigenous peoples, via the use of their traditional knowledge, are able to sustainably manage their lands if they have land security. An example from Peru, shows that with Indigenous tenure, forest disturbance and clearing is significantly reduced (Blackman et al., 2017). Another perspective, this time from Bangladesh, underscores the importance of recognising the historical impacts of colonisation on tenure and how that can affect adaptation practice.

Perspective 3.4
Cultivating Indigenous Community Engagement in Climate Change
Adaptation in Bangladesh
Md. Masud-All-Kamal

Bangladesh is a highly climate-vulnerable country, but all people are not going to be affected equally. People who are marginalised, living close to the natural environment and dependent on climate-sensitive livelihoods are more vulnerable to climate change than others. Bangladesh has over 54 Indigenous groups comprising about three million people, residing mainly in both flatland and hilly forested areas across the country. The most prominent Indigenous groups inhabit the Chittagong Hill Tracts (CHT) – located in south-eastern Bangladesh – who primarily depend on unique agricultural practices called shifting cultivation, locally known as *jhum*, and the gathering of forest resources. The CHT is the home of over twelve Indigenous communities who have historically faced many socio-political and environmental challenges.

The policies and initiatives of the British colonial administration and subsequent governments taken for the protection and development of Indigenous people have proved to be counterproductive. Such interventions have affected traditional administrative structures and the collective/communal land rights and management practices of the Indigenous peoples, which has resulted in displacement, marginalisation and conflict. These discriminatory state policies have also led to the loss of key cultural features including spirituality, nature relationships and rituals. Climate change has added an extra layer to the vulnerability of Indigenous communities in the CHT. Indigenous peoples in this region, like many other Indigenous populations in South Asia and worldwide, depend on generational and old knowledge of weather patterns, but they are now experiencing the effects of climate change in the form of changing seasonal timing, landslides, droughts and floods. All these factors affect sustainable traditional livelihoods and ways of life.

The Indigenous people of the north-eastern region have always addressed environmental variability and adapted for generations, using their cultural and traditional knowledge to cope with and adapt to changing climatic conditions. For instance, in order to protect agroforestry crops from heavy rainfall, Indigenous people cultivate vegetables before the rainy season and cover seedlings and saplings with bamboo nets and support saplings with bamboo sticks to keep them upright (Rahman & Alam, 2016). However, despite their extensive knowledge and experience with coping with environmental change, Indigenous groups in Bangladesh have been ignored in the process of climate change policy-making and planning for the country. Rather than adopting expert-led analyses and solutions based on scientific methods, planners and implementers need to develop adaptation based on sociocultural values, knowledge and prevailing technologies of its Indigenous peoples (Datta, 2019). Without meaningful participation the adaptive capacity of Indigenous communities in Bangladesh will be compromised and entrench marginalisation and deepen their vulnerability. Therefore, adaptation initiatives for building Indigenous peoples' adaptive capacity must sufficiently consider Indigenous modes of climate change adaptation.

Such planning must also sufficiently consider historical and contextual aspects. Historically, Indigenous peoples have experienced disruption, dislocation and deception through state-driven development interventions. The State may interpret climate change adaptation as a new instrument to control and alienate Indigenous groups from their traditional means of livelihoods. For Indigenous communities in Bangladesh, the management of environmental climate variability and change is also intertwined with their rights to the land and forest resources, as well as their rights to utilise their place-based knowledge. The customary rights to forestlands and other natural resources not only act as buffer against diverse stresses, but also preserve their spiritual and cultural values acquired through experience and observations and passed down through stories, apprenticeship and practice. Indigenous adaptation knowledge and practices in Bangladesh are dynamic, cumulative and flexible.

Therefore, any endeavours to enhance the adaptive capacity (or resilience) of Indigenous peoples to climate change in Bangladesh must also uphold their traditional ownership, tenure and management of agricultural and forestlands. In turn, this will strengthen their capacities to adapt to uncertainties including those associated with climate change. Bottom-up approaches can enable critical and culturally sensitive approaches that will document and integrate Indigenous understandings in climate change adaptation policy, planning and programs. Such tenure rights would enable the Indigenous communities of Bangladesh to be recognised and incorporated into climate adaptation policy and practices, to protect both nature and traditional livelihood practices (Photo 3.3).

Photo 3.3 Indigenous people in the Chittagong Hill Tracts use their Indigenous knowledge in traditional shifting cultivation to adapt to climate change. (Credit: Masud Kamal)

Tenure security is also a driver when considering migration as an adaptation option for Indigenous peoples. Often touted as a pathway solution to the catastrophic effects of sea level rise and flooding (amongst other impacts) on small Indigenous communities, it is in fact not a preferred option.

An example, drawn from the Pacific is illustrative. In this case, the normalisation of a discourse around climate change and sea level rise in the Pacific has led to international acceptance that there is very little hope for Pacific Atoll countries. It is believed that Pacific Atoll countries in fact face finite futures and hence must migrate/leave their tribal homes (Barnett, 2017). This in turn has led to the implementation of various forms of policy where adaptation is seen as "palliative care", the step taken just prior to mass migration.

Yet, in the Pacific itself, despite the normalisation of what is a 'loss' discourse, communities are being creative and are simply refusing outright to accept the normalisation process. In a study of this concept in Tuvalu, rather than emphasise a loss discourse (and the argument that migration is inevitable and should be accepted), Barnett (2017) argues for a more creative/hopeful delivery of adaptation. One that seeks to ask the Pacific Islanders what *they* consider is intolerable and what they think would advance an agenda of proactive adaptation.

In answering this question, we find that in the Pacific, Indigenous peoples resist being characterised as vulnerable and forced to relocate; it is **not** perceived by them as the 'only' option. Rather than relocate for example, local peoples in Fiji, Vanuatu and the Solomon Islands have used replacement housing and re-established patterns of self-sufficiency to adapt. In Papua New Guinea, a wide range of sea level rise options are being implemented, instead of asking people to move elsewhere (Damon, 2018). As Bryant-Tokalau (2018, xiii) in a case study of the Solomon Islands, Vanuatu and Fiji notes:

Although the Islands of the Pacific are constantly portrayed as victims in the face of climate change, and indeed are facing many major impacts with increasingly strong storms and rising sea levels, they also have many ways of facing up to the changes. Pacific Islanders have much to teach other nations about resilience and coping and are determined to use their knowledge to maintain ancient traditional practices….far from what is portrayed in the media, islanders and their countries, are not always as vulnerable as they may appear, and have, in the past, the ability to survive in the face of environmental changes.

Further, in moving from one area to another, Indigenous people may face additional language and cultural barriers, housing issues, and discrimination when seeking employment. This challenge has been faced by many Native Alaskans as they have been forced to relocate from their villages into cities due to climate change.

Tenure security then has many resonances in the context of adaptation: being able to ensure stable tenure will assist Indigenous groups to be protected from the sharper impacts of the external globalising stressors described above and allow them a voice to make decisions about and to independently manage disruptors like illegal logging, mining or land appropriations from government. Security of land/sea tenure enables Indigenous groups to build social capital and adaptive capacity by mandating their right to make decisions about what happens in their territories, and provides sources of livelihood and sustenance, which can be a buffer against climate impacts and the need to migrate.

Relationship with Government

Underpinning many of the issues around tenure, resource extraction and building effective adaptation is the relationship Indigenous peoples have with their respective governments. Governments at all levels both support and inhibit Indigenous actions on climate change, with many policy failures typically creating maladaptation and welfare dependence (Son & Kingsbury, 2020). In Vietnam for example, for the Tay, Dao, and Hmong ethnic minorities the Northern Mountainous Region (NMR), the implementation of their local knowledge for climate adaptation remained conditional on institutions and policymakers at the local, regional, and central levels (Son & Kingsbury, 2020). In colonised countries such as the USA, Canada, Australia and New Zealand, government relationships with their First Peoples remains fraught with ongoing tension and further, via the imposition of Western legal systems and they too often curtail the use of Indigenous knowledge so they are not as visible or recognised (Barry & Porter, 2012). Hierarchical governance models are not generally inclusive of Indigenous voices, or cultural modes of governance.

Yet governments have an opportunity to support Indigenous leadership and governance. As Irlbacher-Fox and MacNeill (2020, 271) note "the best climate change adaptation strategy is for governments (and voters) to support Indigenous governance of climate change strategies for their communities and territories, ensuring the provision of resources needed to accomplish targeted outcomes and goals". Collaborative initiatives that are localised and decentralised, ones built between Indigenous communities, state governments, and scientists that intersect at various scales and connect local, national, and international scales is one way forward (Brugnach et al., 2017; Bunce, 2019).

Gender

As much as gender significantly influences how climate change is experienced within Indigenous cultures, so too is it a powerful influencer within Indigenous adaptation, where the deployment of Indigenous knowledge into adaptation can also reflect gender nuances and reinforce key gender roles within different cultures. In many cases these nuances mean that Indigenous women are disproportionately and negatively affected. Hence, the status and also role of Indigenous women in adapting to climate change is important and must be considered. As Jennifer Morris, President of Conservation International notes:

> Indigenous women are disproportionately affected by the climate crisis and they are powerful agents in the fight to halt it… A critical step to protecting nature, to protecting the planet, is elevating the rights and roles of the world's Indigenous peoples, especially women (Price, 2019, 1).

Some adaptation programs are already addressing this issue: in the Amazonia region, the NGO Conservation International works together with the Women's Council of the Coordinator of the Indigenous Organizations of the Amazon Basin (COICA) to build the capacity of, and to empower indigenous women to be conservationists. In Maa, Kenya, women are leading the way in adapting to change by implementing a number of adaptations such as the use of alternative energies, the formation of small economic groups to rotate money, supporting others via village savings and loan associations and making beads for sale.

In some cases, women's attempts to manage for climate change has caused a revitalisation of traditional knowledge. As a woman from Yap Island in the Pacific notes, "climate change is not something new to us, it's always been there and we have ways to fight climate change, there are traditional techniques we can use" (woman from Yap, cited in Mcleod et al., 2018, 181). Other examples of female led adaptation in the Pacific include the establishment of the Palau Community College Cooperative Research and Extension which is experimenting with planting types of salt tolerant taro. Women in Yap are planting palms in flooded taro patches to provide material for weaving and protection from flooding, while in the Marshall Islands, women are braiding Pandanus leaves to guide rainwater into storage containers and plant native plants to reduce coastal erosion and flooding. While Pacific Island women are keepers of important knowledge, their gender and existing power structures mean that their role is under-valued and under-represented. Perspective 3.5 highlights an example of this in practice in Fiji and illustrates the role gender plays to both help and hinder Indigenous adaptation.

Perspective 3.5
Valuing Women's Indigenous Knowledge of Resource Management in Fiji
Jasmine Pearson, Karen E McNamara, Roselyn Kumar

The value of Indigenous knowledge for resource management and climate change adaptation has been well-documented over time. What is often overlooked, however, is the gendered nature of Indigenous knowledge with significant differences between women and men in terms of the type, access, use, transmission and preservation of knowledge. Recognising the gendered distinctiveness of these dynamic and living bodies of knowledge is critical for their ongoing survival, an issue increasingly important in the face of climate change and disaster risk (Nalau et al., 2018). Drawing on a study from Bua Province, Vanua Levu Island, Fiji (see Fig. 3.1), we highlight the gender-specific knowledge held by *iTaukei* (Indigenous Fijian) women.

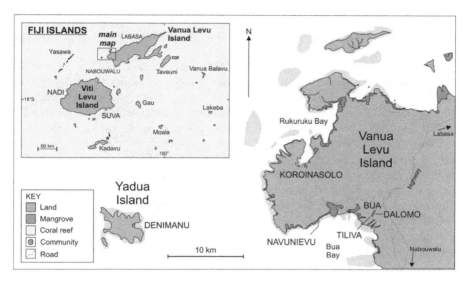

Fig. 3.1 Map of Fiji Islands showing key village sites: Bua, Dalomo, Denimanu, Koroinasolo

Purposive sampling was used to target interview participants regarded as community experts on *dogo* (mangroves). As a result, 85 per cent of participants were women. The findings showed that *iTaukei* women held a substantial body of knowledge on mangrove ecosystems due to their traditional role of foraging and collecting resources such as *qari* (mud crabs). In comparison to men, women held greater knowledge on the value of mangroves for traditional medicine, garlands, artwork, household decorations and firewood (Pearson et al., 2019).

Both women and men identified separate *tabu* areas, community replantation and knowledge sharing through oral traditions as key mangrove management strategies (Pearson et al., 2020). Women were more likely to replant mangroves, and pass on their knowledge to younger generations about managing and taking care of them. The extent of transmitted knowledge retained by younger generations has long been a concern shared by Indigenous women, especially in the age of mobile phones, internet and accumulating stress from climate change. This raises questions about the sustainability of Indigenous knowledge in contexts like rural Fiji (Fache & Pauwels, 2020; McMillen et al., 2017).

Under *qoliqoli* systems (customary marine tenure in Fiji), *tabu* areas are commonly used to temporarily ban or restrict the collection of fish or other marine species from a given area, with permission from the *Ratu* (village Chief) (Aalbersberg et al., 2005; Vuki & Vunisea, 2016). Due to complex power dynamics, Fijian women are often prohibited from decision-making processes regarding *tabu* areas, and general resource management (Ram-Bidesi, 2015; Vunisea, 2007). This is in contrast to the dominantly matrilineal islands of Micronesia where women are empowered to identify gaps in knowledge transmission and largely contribute to community sustainability (Kim, 2020).

Women's Indigenous knowledge is distinct, relevant and valuable, extending from resource management to cultural maintenance and identity (Bryant-Tokalau, 2018). Failure to acknowledge the gendered nature of Indigenous knowledge in climate policy and programs will minimise outcomes, and potentially jeopardise the long-term survival of such knowledge. This is critical for many Pacific Island communities affected by climate change, compromising both their resource base and coping capacities. The latter is being exacerbated by COVID-19 and the growing need for cashless adaptation (Bryant-Tokalau, 2018; Nunn & Kumar, 2019). Understanding Indigenous resilience and unpacking its gendered components is vital for enabling effective and sustainable adaptation solutions. The challenge now is how to promote women's Indigenous knowledge in climate change adaptation without up-ending the customary and cultural dynamics that exist in Pacific Island societies (Photo 3.4).

Photo 3.4 Pacific women adapting to climate change. (Credit: Jasmine Pearson)

Connection and Spirituality

Finally, spirituality is also an important driver – and inhibitor for Indigenous peoples as they try to establish how to adapt to change. This is manifest in two ways. Firstly, Indigenous peoples are spiritually connected to their territories, and believe that this connection needs to be invigorated and upheld as part of responding to climate change:

> In our culture we are encouraged to spend as much time as we can on the land, getting in touch with the Earth, feeling the presence of spirit in the world.We all need to make our way to the land, to recover and to heal the human spirit. We have to feel nature, feel the sun, feel the wind, feel the breeze, feel the rain, and listen to the voice of nature, through the sounds of the animals, sounds of the birds; to listen to messages that come from the winds; to hug the trees, to lay on the land, to feel the love of the earth. We need to listen to the waters in the rapids in the falls, to see the beauty of the land; to see the stars twinkling in the night sky, to see, feel the power of the full moon; to greet the sun at sunrise in gratitude for the blessing of life's gifts, to touch the land with your bare feet. The spirit in the land will guide us, teach us, and ultimately give us our survival (an Anishinabe Elder, cited in Courchene, 2019, 1)

In Maori tradition, every plant, animal and living thing share a cosmological relationship and this relationship means that Māori peoples believe that the wellbeing of one depends on the wellbeing of all. In Peru, where there are 340,000 resident Indigenes, and 42 Amazon ethnicities, the spiritual connection to their territories is so profound that they believe climate change has caused spiritual disruption to their connection to place. As a result, to redress climate impacts on their livelihoods, they have initiated an active recreation of their traditional agricultural practices. This is seen as a fundamental step to ensure that spiritual 'disharmonies', caused by climate change are redressed and so they can re-institute ritual conversations with their lands, and ancestors (Panduro, 2018).

This idea of spiritual 'disharmony' is not uncommon. In the Caribbean coast off Nicaragua, the Miskitu and Garifunas people fear that their spirits are leaving as climate change from their perspective is causing imbalances that cause a loss of spirit values thus leading to less connection with the spirits (Kain, 2018). In Bolivia, the Charzzani, Quechua and Kallawaya nations who live in the Andes, implement adaptive strategies all the time, yet modern technologies and factors like school attendance for their children compromise the practice of their spiritual rituals (Vidaurre de la Riva et al., 2013).

However, the other way in which spirituality becomes an active agent in Indigenous adaptation is where it creates a climate of scepticism or a confusion about managing it. In this case, (and often these examples are set within Christian spiritual tradition), the causes of climate change are constructed as 'lessons' or 'punishments' from God (Ford et al., 2020). The Chuuk Islanders in the Pacific for example see climate change as God's plan to resurrect proper Christian behaviour (Hofmann, 2018, 4). It is important to understand that Christianity is both praxis and politics, when located within sea level rise, migration and climate change.

Understanding how different Indigenous groups construct their spiritual connection to place and how they relate it to climate change can thus inform the development of productive adaptive partnerships, ones that can align with cultural and spiritual connections to place (Kempf, 2020).

Summary

This chapter presents an overview of some of the ways in which Indigenous peoples are responding to climate change: they not only illustrate examples of the diversity of actions being taken but also the challenges in developing them. Nonetheless, Indigenous peoples, whether colonised or not, all face similar threats to their livelihoods, culture, and knowledge, and have drawn on all these elements to build and lead their own futures. It is an exhibition of unparalleled place based yet collective agency. To explore in a deeper way the context and journey of some of these initiatives, the following chapters investigate the drivers for adaptation and lessons learned via three in depth case studies. The first is a case study from North America that analyses how the First Nation peoples of the United States have adapted to climate change on a large scale and provides an insight into Indigenous adaptive governance in action.

References

Aalbersberg, B., Tawake, A., & Parras, T. (2005). Village by village: Recovering Fiji's coastal fisheries. In *World resources, world resource institute* (pp. 144–152).

Afar National Regional State. (2010). *Programme of plan on adaptation to climate change.* Environmental Protection Authority. https://www.semanticscholar.org/paper/Afar-National-Regional-State-Programme-of-Plan-on/fb975755a3aff85a3fdf91319d2f66bbe28ffd17. Accessed 25 Aug 2020.

Allard, C., & Brännström, M. (2021). Girjas reindeer herding community v. Sweden: Analysing the merits of the Girjas case. *Arctic Review, 12*, 56–79.

Axelsson, P. (2010). Abandoning "the other": Statistical enumeration of Swedish Sami, 1700 to 1945 and beyond. *Berichte zur Wissenschaftsgeschichte*, Wiley, *33*(3), 263–279.

Azibo, B. R., & Kimengsi, J. N. (2015). Building an indigenous agro-pastoral adaptation framework to climate change in Sub-Saharan Africa: Experiences from the North West Region of Cameroon. *Procedia Environmental Sciences, 29*, 126–127.

Balehegn, M., Balehey, S., Fu, C., & Liang, W. (2019). Indigenous weather and climate forecasting knowledge among Afar pastoralists of north eastern Ethiopia: Role in adaptation to weather and climate variability. *Pastoralism: Research, Policy and Practice, 9*(1), 1–14.

Barber, M. (2018). Indigenous knowledge, history and environmental change as seen by Yolngu people of blue Mud Bay, northern Australia. In *Indigenous knowledge for climate change assessment and adaptation* (pp. 106–122). Cambridge University Press.

Barnett, J. (2017). The dilemmas of normalising losses from climate change: Towards hope for Pacific atoll countries. *Asia Pacific Viewpoint, 58*(1), 3–13.

Barry, J., & Porter, L. (2012). Indigenous recognition in state-based planning systems: Understanding textual mediation in the contact zone. *Planning Theory, 11*(2), 170–187.

Basdew, M., Jiri, O., & Mafongoya, P. L. (2017). Integration of indigenous and scientific knowledge in climate adaptation in KwaZulu-Natal, South Africa. *Change and Adaptation in Socio-Ecological Systems, 3*(1), 56–67.

Biskupska, N., & Salamanca, A. (2020). *Co-designing climate services to integrate traditional ecological knowledge a case study from Bali, Indonesia.* Stockholm Environment Institute.

Blackman, A., Corral, L., Lima, E. S., & Asner, G. P. (2017). Titling indigenous communities protects forests in the Peruvian Amazon. *Proceedings of the National Academy of Sciences of the United States of America, 114*(16), 4123–4128.

Boillat, S., Serrano, E., Rist, S., & Berkes, F. (2013). The importance of place names in the search for ecosystem-like concepts in indigenous societies: An example from the Bolivian Andes. *Environmental Management, 51*(3), 663–678.

Briones, F. (2018). Climate knowledge of Ch'ol Farmers in Chiapas, Mexico. In Fenly, N. (2018) Forests and the indigenous women of Tuapi: Experiences with promoting food security. In D. Nakashima, I. Krupnik, & J. T. Rubis (Eds.), *Indigenous knowledge for climate change assessment and adaptation.* Cambridge University Press.

Broome, R. (2010). *Aboriginal Australians a history since 1788* (4th ed.). Allen & Unwin.

Brugnach, M., Craps, M., & Dewulf, A. (2017). Including indigenous peoples in climate change mitigation: Addressing issues of scale, knowledge and power. *Climatic Change, 140*(1), 19–32.

Bryant-Tokalau, J. (2018). *Indigenous Pacific approaches to climate change Pacific Island countries.* Palgrave Macmillan.

Bunce, W. (2019). *The Colonisation of thought in contemporary climate change Governance Models.* See https://www.e-ir.info/2019/08/01/the-colonisation-of-thought-in-contemporary-climate-change-governance-models/

Carter, L. (2019). *Indigenous Pacific approaches to climate change in Aotearoa/New Zealand.* Palgrave Macmillan.

Cavalazzi, B., Barbieri, R., Gómez, F., Capaccioni, B., Olsson-Francis, K., Pondrelli, M., & Glamoclija, M. (2019). The Dallol Geothermal Area, Northern Afar (Ethiopia)—An exceptional planetary field analog on Earth. *Astrobiology, 19*(4), 553–578.

Commonwealth of Australia. (2020). *Background paper: Cultural burning practices in Australia,* for the Royal Commission into Natural Disaster Arrangements, Canberra, Australia.

Courchene, D. (2019). Traditional knowledge on climate change. *Cultural Survival,* 11, 2019. https://www.culturalsurvival.org/news/traditional-indigenous-knowledge-climate-change. Accessed 1 Aug 2021.

Damon, F. D. (2018). Seasonal environmental practices and climate fluctuations in island Melanesia: Transformations in a regional system in eastern Papua New Guinea. In D. Nakashima, I. Krupnik, & J. T. Rubis (Eds.), *Indigenous knowledge for climate change assessment and adaptation.* Cambridge University Press.

Datta, R. (2019). *Land-water management and sustainability in Bangladesh: Indigenous practices in the Chittagong Hill Tracts.* Routledge.

De la Riva, M. V., Lindner, A., & Pretzsch, J. (2013). Assessing adaptation–climate change and indigenous livelihood in the Andes of Bolivia. *Journal of Agriculture and Rural Development in the Tropics and Subtropics, 114*(2), 109–122.

Eriksen, S., & Marin, A. (2014). Sustainable adaptation under adverse development? Lessons from Ethiopia. In T. H. Inderberg, S. Eriksen, K. O'Brien, & L. Sygna (Eds.), *Climate change adaptation and development: Transforming paradigms and practices* (pp. 194–215). Routledge.

Fache, E., & Pauwels, S. (2020). Tackling coastal "overfishing" in Fiji: Advocating for indigenous worldview, knowledge, and values to be the backbone of fisheries management strategies. *Maritime Studies, 19*(1), 1–12.

Fenly, N. (2018). Forests and indigenous women in Tuapi: "Return to Auhbi Piakan". In W. Alangui, G. Subido, & J. A. Veasquez (Eds.), *Indigenous women, climate change and forests.* Tebtebba Foundation Press.

Ford, J. D., Smit, B., & Wandel, J. (2006). Vulnerability to climate change in the Arctic: A case study from Arctic Bay, Canada. *Global Environmental Change, 16*(2), 145–160.

Ford, J. D., Smit, B., Wandel, J., Allurut, M., Shappa, K., Ittusarjuat, H., & Qrunnut, K. (2008). Climate change in the Arctic: Current and future vulnerability in two Inuit communities in Canada. *Geographical Journal, 174*(1), 45–62.

Ford, J. D., Stephenson, E., Cunsolo Willox, A., Edge, V., Farahbakhsh, K., Furgal, C., Harper, S., Chatwood, S., Mauro, I., Pearce, T., & Austin, S. (2016). Community-based adaptation research in the Canadian Arctic. *Wiley Interdisciplinary Reviews: Climate Change, 7*(2), 175–191.

Ford, J., King, N., Galappaththi, E., Pearce, T., McDowell, G., & Harper, S. (2020). The resilience of indigenous peoples to environmental change. *One Earth, 2*, 533–543.

Fratkin, E. (2014). Ethiopia's pastoralist policies: Development, displacement and resettlement. *Nomadic Peoples, 18*(1), 94–114.

Gammage, B. (2011). *The biggest estate on earth. How aborigines made Australia.* Allen & Unwin.

Gaski, H. (2008). Nils-Aslak Valkeapää: Indigenous voice and multimedia artist. *AlterNative: An International Journal of Indigenous Peoples, 4*(2), 155–178.

Gearheard, S., Pocernich, M., Stewart, R., Sanguya, J., & Huntington, H. P. (2010). Linking Inuit knowledge and meteorological station observations to understand changing wind patterns at Clyde River, Nunavut. *Climatic Change, 100*(2), 267–294.

Gebeye, B. A. (2016). Unsustain the sustainable: An evaluation of the legal and policy interventions for pastoral development in Ethiopia. *Pastoralism: Research, Policy and Practice, 6*(2), 1–14.

Grey, M. S., Masunungure, C., & Manyani, A. (2020). Integrating local indigenous knowledge to enhance risk reduction and adaptation strategies to drought and climate variability : The plight of smallholder farmers in Chirumhanzu district, Zimbabwe. *Jamba, 12*(1), 1–10.

Hansen, L. I., & Olsen, B. (2006). *Samernas historia fram till 1750.* Liber.

Harvey, C. A., Saborio-Rodríguez, M., Martinez-Rodríguez, M. R., Viguera, B., Chain-Guadarrama, A., Vignola, R., & Alpizar, F. (2018). Climate change impacts and adaptation among smallholder farmers in Central America. *Agriculture & Food Security, 7*(1), 57.

Hill, R., Walsh, F. J., Davies, J., Sparrow, A., Mooney, M., Wise, R. M., & Tengö, M. (2020). Knowledge co-production for Indigenous adaptation pathways: Transform post-colonial articulation complexes to empower local decision-making. *Global Environmental Change, 65*, 102161.

Hofmann, R. (2018). Localizing global climate change in the Pacific. Knowledge and response in Chuuk, Federated States of Micronesia (FSM). *Sociologus, 68*(1), 43–62.

International Indigenous Peoples Forum. (2014). *Statement on Paris agreement.* International Indigenous Peoples Forum.

IPBES. (2019). Summary for policymakers of the global assessment report on biodiversity and ecosystem services of the Intergovernmental Science-Policy Platform on Biodiversity and Ecosystem Services. In S. Díaz, J. Settele, E. S. Brondízio, H. T. Ngo, M. Guèze, J. Agard, A. Arneth, P. Balvanera, K. A. Brauman, S. H. M. Butchart, K. M. A. Chan, L. A. Garibaldi, K. Ichii, J. Liu, S. M. Subramanian, G. F. Midgley, P. Miloslavich, Z. Molnár, D. Obura, A. Pfaff, S. Polasky, A. Purvis, J. Razzaque, B. Reyers, R. R. Chowdhury, Y. J. Shin, I. J. Visseren-Hamakers, K. J. Willis, & C. N. Zayas (Eds.), *Intergovernmental Science-Policy Platform On Biodiversity And Ecosystem Services (IPBES).* IPBES secretariat. 56 pages. https://doi.org/10.5281/zenodo.3553579

IPCC. (2019). In P. R. Shukla, J. Skea, E. C. Buendia, V. Masson-Delmotte, H.-O. Pörtner, D. C. Roberts, P. Zhai, R. Slade, S. Connors, R. van Diemen, M. Ferrat, E. Haughey, S. Luz, S. Neogi, M. Pathak, J. Petzold, J. P. Pereira, P. Vyas, E. Huntley, K. Kissick, M. Belkacemi, & J. Malley (Eds.), *Climate Change and Land: an IPCC special report on climate change, desertification, land degradation, sustainable land management, food security, and greenhouse gas fluxes in terrestrial ecosystems.* IPCC.

Irlbacher-Fox, S., & MacNeill, R. (2020). Indigenous governance is an adaptive climate change strategy. *Northern Review (Whitehorse), 49*, 271–275.

Kain, M. C. (2018). The spirits are leaving: Adaptation and the indigenous peoples of the Caribbean coast of Nicaragua. In D. Nakashima, I. Krupnik, & J. T. Rubis (Eds.), *Indigenous knowledge for climate change assessment and adaptation* (pp. 188–196). Cambridge University Press.

Kalanda-Joshua, M., Ngongondo, C., Chipeta, L., & Mpembeka, F. (2011). Integrating indigenous knowledge with conventional science: Enhancing localised climate and weather forecasts in Nessa, Mulanje, Malawi. *Physics and Chemistry of the Earth, 36*(14–15), 996–1003.

Kempf, W. (2020). Introduction: Climate change and Pacific Christianities. *Anthropological Forum, 30*(3), 215–232.

Kim, M. M. (2020). Nesor annim, niteikapar (good morning, cardinal honeyeater): Indigenous reflections on Micronesian women and the environment. *Contemporary Pacific, 32*(1), 147–163.

Kronik, J., & Verner, D. (2010). *Indigenous peoples and climate change in Latin America and the Caribbean*. The World Bank.

Laidler, G. J. (2006). Inuit and scientific perspectives on the relationship between sea ice and climate change: The ideal complement? *Climatic Change, 78*(2–4), 407–444.

Laidler, G. J., Ford, J. D., Gough, W. A., Ikummaq, T., Gagnon, A. S., Kowal, S., Qrunnut, K., & Irngaut, C. (2009). Travelling and hunting in a changing Arctic: Assessing Inuit vulnerability to sea ice change in Igloolik, Nunavut. *Climatic Change, 94*(3–4), 363–397.

Lantto, P., & Mörkenstam, U. (2015). Action, organisation and confrontation: Strategies of the Sámi movement in Sweden during the twentieth century. In M. Berg-Nordlie, J. Saglie, & A. Sullivan (Eds.), *Indigenous politics: Institutions, representation, mobilisation* (pp. 135–163). ECPR Press.

Leduc, T. B. (2007). Sila dialogues on climate change: Inuit wisdom for a cross-cultural interdisciplinarity. *Climatic Change, 85*(3–4), 237–250.

Magnan, A. K., Schipper, E. L. F., Burkett, M., Bharwani, S., Burton, I., Eriksen, S., & Ziervogel, G. (2016). Addressing the risk of maladaptation to climate change. *Wiley Interdisciplinary Reviews: Climate Change, 7*(5), 646–665.

Makondo, C. C., & Thomas, D. S. (2018). Climate change adaptation: Linking indigenous knowledge with western science for effective adaptation. *Environmental Science & Policy, 88*, 83–91.

Marin, A. (2010). Riders under storms: Contributions of nomadic herders' observations to analysing climate change in Mongolia. *Global Environmental Change, 20*(1), 162–176.

Mathiesen, S., Bongo, M., Burgess, P., Corell, R., Degteva, A., Eira, I., & Vikhamar-Schuler, D. (2018). Indigenous reindeer herding and adaptation to new hazards in the Arctic. In D. Nakashima, I. Krupnik, & J. T. Rubis (Eds.), *Indigenous knowledge for climate change assessment and adaptation* (pp. 198–213). Cambridge University Press.

McDowell, J. Z., & Hess, J. (2012). Accessing adaptation: Multiple stressors on livelihoods in the Bolivian highlands under a changing climate. *Global Environmental Change, 22*(2), 342–352.

Mcleod, E., Arora-Jonsson, S., Masuda, Y. J., Bruton-Adams, M., Emaurois, C. O., Gorong, B., & Whitford, L. (2018). Raising the voices of Pacific Island women to inform climate adaptation policies. *Marine Policy, 93*, 178–185.

McMillen, H., Ticktin, T., & Springer, H. K. (2017). The future is behind us: Traditional ecological knowledge and resilience over time on Hawai'i Island. *Regional Environmental Change, 17*(2), 579–592.

Müller-Mahn, D., Rettberg, S., & Getachew, G. (2010). Pathways and dead ends of pastoral development among the Afar and Karrayu in Ethiopia. *The European Journal of Development Research, 22*(5), 660–677.

Nalau, J., Becken, S., Schliephack, J., Parsons, M., Brown, C., & Mackey, B. (2018). The role of indigenous and traditional knowledge in ecosystem-based adaptation: A review of the literature and case studies from the Pacific Islands. *Weather Climate and Society, 10*(4), 851–865.

Napogbong, L. A., Ahmed, A., & Derbile, E. K. (2021). Fulani herders and indigenous strategies of climate change adaptation in Kpongu community, North-Western Ghana: Implications for adaptation planning. *Climate and Development, 13*(3), 201–214.

Nichols, T., Berkes, F., Jolly, D., Snow, N. B., Community, T., & of Sachs Harbour. (2004). Climate change and sea ice: Local observations from the Canadian Western Arctic. *Arctic, 57*(1), 68–79.

Nickels, S., Furgal, C., Buell, M. & Moquin, H. (2006). *Unikkaaqatigiit – putting the human face on climate change: Perspectives from Inuit of Canada*. Ottawa, Inuit Tapiriit Kanatami, Nasivvik Centre for Inuit Health and Changing Environments, Université Laval and Ajunnginiq Centre (National Aboriginal Health Organization).

NMA (Ethiopian National Meteorological Agency). (2007). *Climate Change National Adaptation Programme of Action (NAPA) of Ethiopia*. Addis Ababa, Ethiopia. From: https://unfccc.int/resource/docs/napa/eth01.pdf. Accessed 29 Aug 2019.

Nunn, P. D., & Kumar, R. (2019). Cashless adaptation to climate change in developing countries: Unwelcome yet unavoidable? *One Earth, 1*(1), 31–34.

Nzeadibe, T. C., Egbule, C. L., Chukwuone, N. A., Agwu, A. E., & Agu, V. C. (2012). Indigenous innovations for climate change adaptation in the Niger Delta region of Nigeria. *Environment, Development and Sustainability, 14*(6), 901–914.

Oviedo, A. F., Mitraud, S., McGrath, D. G., & Bursztyn, M. (2016). Implementing climate variability adaptation at the community level in the Amazon floodplain. *Environmental Science & Policy, 63*, 151–160.

Panduro, R. (2018). Peasants of the Amazonian-Andes and their conversations with climate change in the San Martín Region. In D. Nakashima, I. Krupnik, & J. T. Rubis (Eds.), *Indigenous knowledge for climate change assessment and adaptation* (pp. 254–264). Cambridge University Press.

Pearce, T., Ford, J., Willox, A. C., & Smit, B. (2015). Inuit traditional ecological knowledge (TEK), subsistence hunting and adaptation to climate change in the Canadian Arctic. *Arctic, 68*(2), 233–245.

Pearson, J., McNamara, K. E., & Nunn, P. D. (2019). Gender-specific perspectives of mangrove ecosystem services: Case study from Bua Province, Fiji Islands. *Ecosystem Services, 38*, 1–10.

Pearson, J., McNamara, K. E., & Nunn, P. D. (2020). *iTaukei* ways of knowing and managing mangroves for ecosystem-based adaptation. In W. L. Filho (Ed.), *Managing climate change adaptation in the Pacific Region* (pp. 105–127). Springer.

Peña, G., Areola, Q., Tanael, C., Paler, F., Cortez, D., Librero, A., Flores, J., Viajedor, J., Lugatiman, J., Talingting, R., & Vedra, S. (2017). Adaptive measures to climate change among the Higaonon communities in Naawan and Initao, Misamis Oriental, Mindanao, Philippines. *American Journal of Social Sciences, Arts and Literature, 4*(1), 1–7.

Pennesi, K., Arokium, J., & McBean, G. (2012). Integrating local and scientific weather knowledge as a strategy for adaptation to climate change in the Arctic. *Mitigation and Adaptation Strategies for Global Change, 17*(8), 897–922.

Pietikäinen, S., Huss, L., Laihiala-Kankainen, S., Aikio-Puoskari, U., & Lane, P. (2010). Regulating multilingualism in the North Calotte: The case of Kven, Meänkieli and Sámi Languages. *Acta Borealia, 27*(1), 1–23.

Price, E. (2019). *Meet 3 Indigenous women fighting to save the planet, in Conservation International Website*, https://www.conservation.org/blog/meet-3-indigenous-women-fighting-to-save-the-planet. Accessed 1 May 2021.

Rahman, M., & Alam, K. (2016). Forest dependent indigenous communities' perception and adaptation to climate change through local knowledge in the protected area—A Bangladesh case study. *Climate, 4*(12), 1–25.

Ram-Bidesi, V. (2015). Recognizing the role of women in supporting marine stewardship in the Pacific Islands. *Marine Policy, 59*, 1–8.

Reid, M. G., Hamilton, C., Reid, S. K., Trousdale, W., Hill, C., Turner, N., Picard, C. R., Lamontagne, C., & Matthews, H. D. (2014). Indigenous climate change adaptation planning using a values-focused approach: A case study with the Gitga'at nation. *Journal of Ethnobiology, 34*(3), 401–424.

Reimerson, E. (2015). *Nature, culture, rights: Exploring space for indigenous agency in protected area discourses*. PhD thesis, Umeå University.

Reisinger, A., Kitching, F., Chiew, L., Hughes, P., Newton, S., Schuster, A., Tait, & Whetton, P. (2014). Australasia. In V. R. Barros, C. B. Field, D. J. Dokken, M. D. Mastrandrea, K. J. Mach, T. E. Bilir, M. Chatterjee, K. L. Ebi, Y. O. Estrada, R. C. Genova, B. Girma, E. S. Kissel, A. N. Levy, S. MacCracken, P. R. Mastrandrea, & L. L. White (Eds.), *Climate change 2014: Impacts, adaptation, and vulnerability: Part B: Regional aspects: Contribution of working group II to the fifth assessment report of the intergovernmental panel on climate change*. Cambridge University Press.

REXSAC (Resource Extraction and Sustainable Arctic Communities) (2021). *Nordic centres of Excellence in Arctic Research*, at https://www.nordforsk.org/projects/resource-extraction-and-sustainable-arctic-communities-rexsac. Accessed 1 Aug 2021.

Riedlinger, D., & Berkes, F. (2001). Contributions of traditional knowledge to understanding climate change in the Canadian Arctic. *Polar Record, 37*(203), 315–328.

Roué, M. (2018). 'Normal' catastrophes or harbinger of climate change? Reindeer-herding Sami facing dire winters in Northern Sweden. In D. Nakashima, I. Krupnik, & J. T. Rubis (Eds.), *Indigenous knowledge for climate change assessment and adaptation* (pp. 229–256). Cambridge University Press.

Saboohi, R., Barani, H., Khodagholi, M., Sarvestani, A. A., & Tahmasebi, A. (2019). Nomads' indigenous knowledge and their adaptation to climate changes in Semirom City in Central Iran. *Theoretical and Applied Climatology, 137*(1–2), 1377–1384.

Sakakibara, C. (2018). Climate change, whaling tradition and cultural survival among the Iñupiat of Arctic Alaska. In D. Nakashima, I. Krupnik, & J. T. Rubis (Eds.), *Indigenous knowledge for climate change assessment and adaptation* (pp. 265–279). Cambridge University Press.

Samer. (2021). *Samerna i siffror*. http://samer.se/1536. Accessed 18 June 2021.

Sánchez-Cortés, M. S., & Chavero, E. L. (2018). Local responses to variability and climate change by Zoque indigenous communities in Chiapas, Mexico. In *Indigenous knowledge for climate change assessment and adaptation* (pp. 75–83). Cambridge University Press.

Sayles, J. S., & Mulrennan, M. E. (2010). Securing a future: Cree hunters' resistance and flexibility to environmental changes, Wemindji, James Bay. *Ecology and Society, 15*(4), 1–21.

Schmidt, M., & Pearson, O. (2016). Pastoral livelihoods under pressure: Ecological, political and socioeconomic transitions in Afar (Ethiopia). *Journal of Arid Environments, 124*, 22–30.

Sehlin MacNeil, K. (2015). Shafted. A case of cultural and structural violence in the power relations between a Sami Community and a mining company in Northern Sweden. *Ethnologia Scandinavica: A Journal for Nordic Ethnology, 45*, 73–88.

Sehlin MacNeil, K. (2017). *Extractive violence on indigenous country: Sami and aboriginal views on conflicts and power relations with extractive industries*. PhD thesis, Umeå University, Sweden.

Son, H., & Kingsbury, A. (2020). Community adaptation and climate change in the Northern Mountainous Region of Vietnam: A case study of ethnic minority people in Bac Kan Province. *Asian Geographer, 37*(1), 33–51.

Son, H. N., Chi, D. T. L., & Kingsbury, A. (2019). Indigenous knowledge and climate change adaptation of ethnic minorities in the mountainous regions of Vietnam: A case study of the Yao people in Bac Kan Province. *Agricultural Systems, 176*, 102683.

Tilahun, M., Angassa, A., & Abebe, A. (2017). Community-based knowledge towards rangeland condition, climate change, and adaptation strategies: The case of Afar pastoralists. *Ecological Processes, 6*(29), 1–13.

TSRA. (2014, July). *Torres strait climate change strategy 2014–2018*. Report prepared by the Land and Sea Management Unit, Torres Strait Regional Authority, 36p.

UNFCC. (2009). *The anchorage declaration*. UNFCC.

Vidaurre de la Riva, M., Lindner, A., & Pretzsch, J. (2013). Assessing adaptation – Climate change and indigenous livelihood in the Andes of Bolivia. *Journal of Agriculture and Rural Development in the Tropics and Subtropics, 114*(2), 109–122.

Vogt, N., Pinedo-Vasquez, M., Brondízio, E. S., Rabelo, F. G., Fernandes, K., Almeida, O., Riveiro, S., Deadman, P. J., & Dou, Y. (2016). Local ecological knowledge and incremental adaptation to changing flood patterns in the Amazon delta. *Sustainability Science, 11*(4), 611–623.

Vuki, V. C., & Vunisea, A. (2016). Gender issues in culture, agriculture and fisheries in Fiji. *SPC Women in Fisheries Information Bulletin, 27*, 15–18.

Vunisea, A. (2007). Women's changing participation in the fisheries sector in Pacific Island countries. *SPC Women in Fisheries Information Bulletin, 16*, 24–27.

Weatherhead, E., Gearheard, S., & Barry, R. G. (2010). Changes in weather persistence: Insight from Inuit knowledge. *Global Environmental Change, 20*(3), 523–528.

Wheeler, H. C., Danielsen, F., Fidel, M., Hausner, V., Horstkotte, T., Johnson, N., … Root-Bernstein, M. (2020). The need for transformative changes in the use of indigenous knowledge along with science for environmental decision-making in the Arctic. *People and Nature (Hoboken, N.J.), 2*(3), 544–556.

Xu, J., & Grumbine, R. E. (2014). Integrating local hybrid knowledge and state support for climate change adaptation in the Asian Highlands. *Climatic Change, 124*(1), 93–104.

Chapter 4
Tribal Capacity Building and Adaptation Planning: The United States

Ann Marie Chischilly, Nicolette Cooley, Karen Cozzetto, and Julie Maldonado

Introduction

This chapter, the first of our three focus chapters offers a detailed perspective on the climate change story in the United States and what one particular institution, the Institute for Tribal Environmental Professionals (ITEP) at Northern Arizona University, has been working on to address climate-related concerns/issues. The chapter highlights not only the human face of climate change, but the ways in which previous colonial trauma exacerbates its impacts. Further, it explores how complex climate governance can be, when engaging with Indigenous needs and aspirations, not just when addressing impacts but in working out how to have a voice at the decision-making table.

Adaptation and Surviving Trauma

The concept of adaptation for American Indians is not new and was forced on the tribes of the United States (US) as they worked out how to survive upon colonial settlement pre-1491. American Indian tribal nations have since endured centuries of genocidal type treatment from non-Indigenous peoples and the US federal government.[1]

---

[1] The term "American Indians", "tribes" and "tribal nations" are used interchangeably unless referencing a specific group or a specific region.

A. M. Chischilly
Tribal Environmental Professionals, Northern Arizona University, Flagstaff, AZ, USA
e-mail: Ann-Marie.Chischilly@nau.edu

N. Cooley (✉) · K. Cozzetto · J. Maldonado
Northern Arizona University, Flagstaff, AZ, USA
e-mail: nikki.cooley@nau.edu; karen.cozzetto@nau.edu

© The Author(s) 2022
M. Nursey-Bray et al., *Old Ways for New Days*, SpringerBriefs in Climate Studies, https://doi.org/10.1007/978-3-030-97826-6_4

Beginning in the late 1500s and 1600s, there have been millions of Indigenous deaths across the Americas primarily caused by disease and European-led massacres (Koch et al., 2019).

In the United States today there are more than 5.2 million American Indian and Alaskan Native people from across 573 federally recognised nations. There are 375 signed treaties, passed laws, and instituted policies that today continue to shape and define the unique and legal contracts that form the government-to-government relationship between the US and these tribal nations. However, almost all of the treaties were signed or agreed to under duress or compulsion. American Indian tribal nations conceded millions of acres of their homelands for a set of rights which include but are not limited to: guaranteed peace and protection, land reserves or legal land boundaries, preservation of hunting and fishing rights, education, health care, self-government and jurisdiction over their own lands. Those rights have not been implemented to the satisfaction of tribal nations and American Indians continue to fight for visibility, recognition and representation.

These colonial beginnings and their legacy are at the root of many of the challenges that American Indians face in the US today, and now, in relation to climate change. Further, the ongoing refusal to honor treaties means that today American Indians endure a continued traumatic syndrome (CTS). The 573 tribal nations [within the US] hold substantial amounts of land and waterways that are being impacted by climate change including habitat for more than 525 species listed under the Endangered Species Act, and more than 13,000 miles of rivers and 997,000 lakes are located on federally recognised tribal lands (Jantarasami et al., 2018, 578). Climate threats to Indigenous peoples' livelihoods and economies will also affect agriculture, hunting and gathering, fishing, forestry, energy, recreation, tourism enterprises and finally traditional subsistence economies:

> Such economies rely on local natural resources for personal use (such as food, shelter, fuel, clothing, tools, transportation, and arts and crafts) and for trade, barter, or sharing. Climate change threatens these delicately balanced subsistence networks by, for example, changing the patterns of seasonal timing and availability of culturally important species in traditional hunting, gathering, and fishing areas. (Jantarasami et al., 2018, 579)

Related to this is the reality that many First Nation peoples of the United States still have much higher rates of poverty and unemployment compared with the national average (Krogstad, 2014; Norton-Smith et al., 2016, 4) and the combination of both climate impacts and poverty will compound existing inequalities. The Tribal chapter in the National Climate Assessment, also highlights the severity of these impacts, noting that climate change is not only having an impact on many of the 566 federally recognised tribes and other and Indigenous groups in the U.S. but that they are compounded by a range of persistent social and economic problems (Bennett et al., 2014).

However, we argue that American Indians have turned their continued trauma and their reaction to these other issues into a positive energy, one which has become a driver of resilience. Tribes have learned that adaptation is a 'state of mind', an 'unspoken way of life', necessary to endure and thrive. Tribal strength and agency continue through teachings, language, traditional knowledge and culture. There is a tacit understanding among tribal people that despite all the obstacles, tribes will persevere and eventually: to lead. Tribes continue to maintain old traditions today. This is nowhere more evident than in the country wide First Nation led tribal adaptation planning that is underway. The energy and agency of these initiatives represent the diversity of responses and collective effort taken by First Nations groups to address climate issues across the United States. The rest of this chapter provides an overview of and reflection about this example of nationwide First Nation led adaptation program.

Tribal Adaptation Planning

At a national level, one key initiative has been the development of Tribal Climate Change Adaptation Plans (TCCP). The TCCP program is designed to assist tribal professionals and their communities to develop their own Tribal CC Adaptation Plan (CC Plan), an initiative that was boosted in 2011, by the Bureau of Indian Affairs (BIA) Resilience Program, which began to offer funding for training and technical assistance to tribes. Each of the climate change adaptation plans that have been developed are unique, having become, in essence, sacred documents that include each tribes' history, challenges, aspirations, laws, culture, traditional knowledge, data and holistic implementation goals. Divided into specific regions for adaptation planning, much headway has now been made: 50 of the 573 (9%) tribes across the United States now have some form of a climate change adaptation plan or assessment. Planning has occurred across the Northeast and Southeast; Midwest; Northern and Southern Great Plains; Northwest; Southwest regions as well as in Alaska. While this large number makes it difficult to summarise each of the current tribal climate change initiatives, we present a range of examples below.

Northwest Region Tribes

The Northwest Region tribes, known as People of the Salmon, are fishing and hunting peoples and rely heavily on the ocean, rivers and lakes for subsistent lifeways. The Northwest region was one of the last areas to be contacted by non-First Nation peoples, however, once contact occurred, diseases like smallpox killed about 90% of the local tribes (Crosby, 1976). In this region, climate impacts also include temperature changes which are of particular significance because they exacerbate existing stresses on salmon and shellfish populations which has an impact on the economic, spiritual, and cultural health of communities (Norton-Smith et al., 2016).

Other impacts include early snowpack melts (causing flooding), increased wildfires and sea levels, and associated impacts such as storm surges, relocation, ocean acidification, mammal and fish migration, droughts, salt-water intrusion and erosion (DOE, 2015; ITEP website, 2019). However, in this region, tribes have bound together in a powerful union to fight climate change and it was the first in the United States to build climate change plans: the Swinomish Nation was the first tribe to complete a climate change plan in 2010. The range of work undertaken is significant as shown in Box 4.1 below which lists the first swathe of tribal climate adaptation plans to be produced.

Box 4.1 Summary of Tribal Climate Adaptation Plans in the USA

- Swinomish Climate Change Initiative: Climate Adaptation Action Plan;
- Swinomish Climate Change Initiative: Impact Assessment Technical Report;
- Puyallup Tribe of Indians Climate Change Impact Assessment and Adaptation Options – 2016;
- Clearwater River Sub-basin (ID) Climate Change Adaptation Plan;
- Climate Adaptation Plan for the Territories of the Yakama Nation;
- Climate Change Preparedness Plan for the North Olympic Peninsula;
- Confederated Tribes of the Umatilla Indian Reservation Climate Change Vulnerability Assessment;
- Flood and Erosion Hazard Assessment for the Sauk-Suiattle Indian Tribe;
- Forest and Water Climate Adaptation: A Plan for the Nisqually Watershed;
- Jamestown S'Klallam Tribe-Climate Vulnerability Assessment and Adaptation Plan;
- Lummi Nation Climate Change Mitigation and Adaptation Plan: 2016–2026;
- Makah Tribe's Climate Resilience, Adaptation, and Mitigation Planning;
- Nooksack Indian Tribe Natural Resources Climate Change Vulnerability Assessment Indian Tribe Natural Resources Climate Change Vulnerability Assessment
- Shoalwater Bay Indian Tribe 2014 Hazard Mitigation Plan; and
- Stillaguamish Tribe Natural Resources Climate Change Vulnerability Assessment.
- Nez Perce Tribe Climate Change and Community Well-Being Survey: Results and Discussion;
- Shoshone-Bannock Tribes: Climate Change Assessments and Adaptation Plan;
- Upper Snake River Tribes Foundation Climate Change Vulnerability Assessment
- Colville Tribes Natural Resources Climate Change Vulnerability Assessment

North and South Eastern Region Tribes

The North and South Eastern Region is a very large region covering the entire eastern side of the US, however there are less tribes in these regions, due in large part to being the first tribes impacted by contact with the colonisers: "[B]y the late 19th century, fewer than 238,000 Indigenous people remained, a sharp decline from the estimated 5 million to 15 million living North America…" (Fixico, 2018, 127). Most of the tribes in this large region are coastal tribes or close to water bodies and are avid fishers. In the northern region, there are many large mammals that the tribes hunt, including moose and deer. The northern tribes also rely on tapping the maple trees for maple syrup. In this region of the US, the climate change impacts that are projected to occur include: increased temperatures and more frequent and longer heat waves; Atlantic hurricanes will be more frequent at either categories 4 or 5, and sea levels will rise (DOE, 2015).

Tribes in this region are also experiencing the migration of subsistence animals, migration and decline of fowl and marine life, migration of forest lines, sea level rise, ocean acidification and increased precipitation. Two tribal adaptation plans for this region include the Climate Change Adaptation for Akwesasne Saint Regis Mohawk Tribe and the Shinnecock Indian Nation Climate Adaptation Plan.[2] These plans embed practical actions that the tribe can institute to help them adapt to the ongoing and expected climate impacts: they also represent innovative forms of tribal governance.

Midwest Region Tribes

The Midwest Region of the US is in the middle of the country. This region contains all the states that surround the five Great Lakes and borders Canada. The tribes in the region share much of the same history of the eastern tribes with regards to being some of the first tribes to be contacted, however more tribes in this region were able to escape, be relocated, and survive. In this region, the tribes are well regarded hunters, fishers, farmers, and rice harvesters. They are stewards of their forests and rice fields. As in other regions, the climate impacts will include increased temperature, longer and more severe heat waves, increased lake temperatures which will increase incidence of toxic algal blooms, and heavier than average winter and spring precipitation levels which will cause flooding (DOE, 2015). Box 4.2 highlights the range of adaptation plans developed in this region to address climate change.

Box 4.2 Range of Adaptation Option for the Midwest Region Tribes

- 1854 Ceded Territory Including the Bois Forte, Fond du Lac, and Grand Portage Reservations: Climate Change Vulnerability Assessment and Adaptation Plan;
- Bad River Reservation: Seventh Generation Climate Change Monitoring Plan;
- Great Lakes Indian Fish and Wildlife Commission (GLIGWC) Climate Change Vulnerability Assessment: Integrating Scientific and Traditional Ecological Knowledge;
- Fond Du Lac Resource Management – 2008 Integrated Resource Management Plan;
- Match-e-be-nash-she-wish Band of Pottawatomi Indians Climate Change Adaptation Plan;
- Michigan Tribal Climate Change Vulnerability Assessment and Adaptation Planning: Project Report;

---

[2] Please click on this link from the University of Oregon's Tribal Climate Change Project led by Kathy Lyn for examples of many tribal climate adaptation plans: https://tribalclimateguide.uoregon.edu/adaptation-plans

- Resilience Dialogues- Final Synthesis Report Menominee Reservation, USA;
- Mitigwaki idash Nibi: (Our Forests and Water) A Climate Adaptation Plan for the Red Lake Band of Chippewa Indians;
- Red Lake Band of Chippewa Climate Adaptation Plan.

Northern and Southern Great Plains Region Tribes

The Northern and Southern Great Plains region tribes are renowned for their great horsemanship skills and rely heavily on buffalo (bison), fishing, gathering, and farming. This region has vast forest lands, large prairies, and rivers and large water bodies. One of the greatest hardships in this region is its high poverty rate; unemployment is one of the highest in the US at 51.9% (Stebbins & Sauter, 2019), and life expectancy is 57 years old which is 24 years less than for the non-First Nation residents (Walker, 2019). In this region, several climate change impacts are projected including: heavy precipitation leading to flooding, an increase in average temperature, decrease in water availability, drought, more tornados, and intense blizzards. Southern tribes will also face intense Atlantic hurricanes from the Gulf of Mexico as well as sea level rise (DOE, 2015). There are three key plans for this region: (i) the Blackfeet Climate Change Adaptation Plan, (ii) The Climate Change Strategic Plan Confederated Salish and Kootenai Tribes of the Flathead Reservation and the (iii) Ovate Omnicive' Oglala Lakota Plan, which is the official, regional sustainable development plan for the Oglala Sioux Tribe.

Southwest Region Tribal Profiles

The Southwest region is distinguished by the fact that, apart from Alaska, many tribes were dry farmers, gatherers, pastoralists, herdsmen and gatherers. However, the tribes in the region also contended with invasion and disease: they were finally reduced to very small numbers and Rancherias. Tribes in this region are both coastal and inland and face drought intensification, increased wildfires, loss of agriculture and stock (Norton-Smith et al., 2016). Again, notwithstanding this history of loss and invasion, as Box 4.3 shows, the tribes in this region have worked together to produce a wide range of responses to climate challenges including adaptation and mitigation plans, vulnerability assessments and resource management plans (Photo 4.1).

Box 4.3 Tribes Responses to Climate Change in the Southwest Region

- Bear River of the Rohnerville Rancheria Climate Change Mitigation and Adaptation Action Plan 2018;
- Karuk Tribe Department of Natural Resources Eco-Cultural Resources Management Plan;
- Karuk Tribe Climate Vulnerability Assessment Assessing Vulnerabilities from the Increased Frequency of High Severity Fire;
- Karuk Climate Adaptation Plan;
- Navajo Nation Climate-Change Vulnerability Assessment for Priority Wildlife Species;
- Vulnerabilities of Navajo Nation Forests to Climate Change;
- Climate Adaptation Plan for the Navajo Nation;
- Susanville Indian Rancheria: Integrated Resource Management Plan;
- Campo Climate Adaptation Action Plan;
- Climate Change Vulnerability Assessment – Pala Band of Mission Indians; and
- Yurok Tribe Climate Change Adaptation Plan for Water & Aquatic Resources 2014–2018

Photo 4.1 Monument Valley, part of the tribal territories of the Navaho Nation. (Credit Melissa Nursey-Bray)

Factors Facilitating Adaptation Success

This wide range of climate planning across the United States shows collective effort but also captures the unique character of each tribe. Each climate change plan captures the individual tribal history, culture, way of life, and details how each tribe is addressing climate impacts. Each plan is tailored to the specific group and region. The overall goal for these tribal climate change plans is to maintain tribal ways of life on sacred territories. There are a range of support mechanisms in place that enable successful implementation of these plans.

The first of these is tribal leadership. Tribes offer leadership at various levels and some tribal leaders like the President of the National Congress of American Indians (NCAI) and Quinault Tribal President, Fawn Sharp, have committed to support Native Americans to tackle the challenges before them (Browning & Aegerter, 2019). Another US climate change leader is Karen Diver, former President Obama's Special Assistant to Native Americans Affairs and former Chairwoman of the Fond du Lack Band of Lake Superior Chippewa. She worked within the Obama Administration and was appointed to the Tribal Leaders Task Force on Climate Preparedness and Resilience.

Inter-Tribal Consortiums and Organisations (ITC) are other means by which tribes are supported in climate actions. The ITC are coalitions of two or more separate Indian tribes that join together for the purpose of participating in self-governance, including Tribal organisations (LII, 2019). They play a major role in assisting tribes in their climate change planning and lobbying efforts. There are currently about twenty-four Inter-Tribal consortiums in the US (NCAI website, 2019) that meet regularly. Examples of such consortiums include the Affiliated Tribes of the Northwest Indians (ATNI); Inter-Tribal Council of Arizona, Great Plains Tribal Chairman's Association, and the United South and Eastern Tribes.

Another source of support derives from the National Congress of American Indians (NCAI), which, founded in 1944, every year passes a series of resolutions to help tribes on certain issues. In this context, the NCAI has passed many resolutions relevant to climate change, one good example being the Guidance Principles to Address the Impacts of Climate Change (NCAI, 2013), which calls on the United Nations Framework Convention on Climate Change to adopt an agreement that Upholds the Rights on Indigenous Peoples (NCAI, 2013), and support tribes to respond to Federal policies and actions to address climate change. In 2017, the tribes voted to unanimously adopt Resolution MOH-17-053 entitled, "Continued Support for the Paris Climate Agreement and Action to Address Climate Change". The Resolution seeks commitments that support initiatives that will help reduce greenhouse gas emissions, promote climate-resiliency and challenges all tribal nations to uphold the Paris Accord (NCAI Website, 2019).

Outside of these plans, there are a few other tribal climate initiatives worth noting that offer additional support to tribes seeking to adapt to climate change. For example, the Pala Band of Mission Indians has developed a Tribal Climate Health website containing news items, and is a searchable clearinghouse of over 500 resources related to health and climate change, and provides tools to assist tribes with integrating health assessment, prioritisation, and tracking into Climate Action Plans. The Tribe also hosts free online webinar training that focusses on the intersection between climate change and tribal health. Another example is the establishment of the Native Youth Community Adaptation and Leadership Congress which is working to build lasting relationships between several federal agencies and native youth, to deepen understanding and build more resilient, adaptable communities. Through this student-led program, native youth gain leadership skills, knowledge, and support to implement meaningful projects in their home communities and create inter-generational change. Finally, the Climate Science Alliance has partnered with the Pala Band of Mission Indians to expand their Climate Kids program into tribal communities in and around San Diego County and throughout the U.S., and *ClimateKids-Tribes* offers educational outreach activities, field trips, climate challenges, and a popular educational resource called the *Traveling Trunks*.

Barriers to Adaptation

Despite the dynamism of tribal climate action, First Nations across the US face many difficulties that inhibit the ongoing success and implementation of these initiatives. Firstly, tribal leaders are inundated with the need to respond to and manage many other urgent priorities such as crime, lack of housing, poverty, substance abuse, suicide, domestic violence and unemployment. As Bennett et al. (2014), convening lead authors for the NCA3 in the 2014 Indigenous Peoples chapter state:

> Indigenous people communities [or tribes] already face many socio-economic challenges, even before overlaying climate change impacts on them and that climate change impacts will exacerbate these challenges, affecting native communities' ability to hunt and gather traditional foods, perform ceremonies, even travel. Indigenous Peoples are starting to see a change in how [they] interpret the environment around [them] (NCA3, 2014).

These historical stressors compound already difficult economic and social conditions within tribal communities.

Another barrier for tribal nations is the limited tribal staff capacity and high turnover rate of tribal employees. Tribes vary drastically in population size; thus, their tribal government and departments also vary in size. The larger tribes may have over fifty employees, however, a large majority of tribes have less than five in their department. Having a limited tribal staff capacity often leads to over extension of the staff which may lead to employee burn out and having the staff member leave their position.

Time is also often a factor: there is too often limited time and funding to draft a Climate Change Plan and many of the tribes do not have an employee fully dedicated to working on climate change issues (Wotkyns & Gonzalez-Maddux, 2014). According to Rachel Novak, Tribal Resilience Program Coordinator from the Bureau of Indian Affairs (BIA), "about 10 per cent [~57 of 573] of federally recognised tribes have a [CC Plan] drafted, but the remaining tribes, still over 500, are at various points in the process, from implementing adaptation plans to assessing possible impacts, to not having begun. Time and resources are usually the biggest barriers to creating a plan" (Ma, 2018). Further, funding for full-time employees is very difficult to secure and maintain. Most tribal environmental offices designate a portion of an employee's time (i.e. part-time) to work on Tribal Resilience Program issues, but overall the lack of full-time commitment lends itself to non-completion of the climate change plans. The first and greatest need is provision of adequate funding to support climate change initiatives:

> Funding limitations are often identified as a barrier to the planning or implementation to climate adaptation or mitigation actions, which suggests that increase economic revenues could create opportunities for tribes to choose to pursue climate actions (Jantarasami et al., 2018 578)

Another huge challenge is working out how to acknowledge and include traditional tribal knowledge into adaptation planning. Traditional Knowledge (TK) as a term does not have a universal definition and every tribe and tribal member has the right to define their TK in their own way. It is precisely due to this lack of a precise definition that its use can become a complicated venture and tribes have fought for the right to use – or not use – their TK when developing their climate change plans. However, this ambition is complicated by the fact that the information the tribe wants to incorporate in their climate change plan could be endangered if published in federally grant funded documents. Safety concerns became apparent for example, when tribes began developing their climate change plans in a holistic manner and into which they included knowledge about prayers, songs, sacred locations, sacred ceremonies, sacred places, medicinal plants, location of plants, use of plants, to name a few. Tribes became concerned that the incorporation of their TK could expose them to theft and general misuse of TK (Wotkyns & Gonzalez-Maddux, 2014).

Related to this is the issue of reciprocity: in 2015 while climate change was starting to be addressed in the US, tribes were not being included in many of the federal initiatives and funding – and government employees did not understand nor respect tribal traditional knowledge as they did Western science/knowledge. Funding for federal grants was consistently awarded to organisations that used western science as a basis for research, yet for many tribes, western science was not only not available but also not culturally compatible. At that time, federal and state agencies did not know what TK was or why it was important to tribes.

Further, climate data collection on tribal lands is problematic as many nations do not trust federal and state agencies due to historical colonisation. Tribes are sovereign and can dictate who is allowed on their lands; yet for many decades, tribes simply did not allow state officials or researchers on their lands to monitor their environment. As many tribal nations also do not have the funding to monitor and collect data on their lands, climate data that may be readily available for states, just does not exist for tribal nations. When tribes began applying for federal funding to support climate change initiatives, they opted to use their TK in lieu of western science because it was available and culturally appropriate for their Plans. Tribes wanted to use their own knowledge from their Elders and citizens. One classic example is an instance where the intent is to monitor a stream. A tribe would want to use an Elder's account of the stream. This Elder who had lived on that stream for 50 years could relate as much or more about the stream, equivalent to the data that would have been collected in a scientific manner. However, when tribes applied for funding from the federal government, their grant applications were being rejected as they were seen to be based on non-western (and hence invalid) knowledge and methods.

It took a collective effort by many tribal experts in the climate change arena to educate the federal and state governments about the definition of TK and its value. One noted expert stated that,

> It is detrimental for the federal government to exclude tribes in climate-change initiatives because long histories of adaptation in response to colonialism, genocide, forced relocation, and climatic events have provided tribes with extensive experience with resistance, resilience, and adaptation (Warner, 2015).

It was at this time that the *Guidelines for Use of Traditional Knowledge in Climate Change Initiatives* (Guidelines) (CTKW, 2014) were developed and these lay the foundation of and principles for partnerships between tribes/Indigenous groups and the federal government.

Finally, the idea of relocation – as a potential adaptation to climate change remains contentious, especially since the colonial history of the forced relocation of US tribes is one of trauma and has not been forgotten:

> The [forced] dispossession and displace of [tribes] from their land [which] began during European exploration and colonization of North America in the 16th century. In those 500 years, over a billion acres of land were coerced from tribes under threat of violence, encroachment, and catastrophic epidemic (Burkett et al., 2017)

Most of those tribes never returned to their original territories and were never compensated for the theft of their lands. The suggestion then that tribes may need to relocate again, this time due to climate change impacts, is a highly traumatic one and yet one of the high priority considerations in climate change planning for tribes:

> Most of the people currently dealing with climate change-induced relocation are Native Americans and Alaska Natives (NAAN) living close to coastal resources... NAAN are vulnerable to sea level rise (Burkett et al., 2017)

Of the 573 tribal nations, 13–17 tribal nations are urgently threatened by sea rise and thus actively engaged in discussions about the feasibility of relocation. The first issue that tribes must contend with is deciding whether or not to move. A collective tribal decision to move from ancestral lands is fundamentally heart breaking. Tribes are place based peoples thus relocation would abruptly split every aspect of their lifeways. Julie Maldonado who has worked with the Isle de Jean Charles Band of Biloxi-Chitimacha-Choctow Indians (IDJC) of southern Louisiana points out that community relocation efforts are

> ...not just about moving a house from X to Y. Rather, it's crucial that tribes, particularly when they decide to move their entire community, are able to choose a safe location that is close enough to their original home that they can still access their natural and cultural resources and sites of cultural importance (cited in Dermansky, 2019).

Tribes also need to be able to sustain a sense of and similar structures to their community in ways that culturally align and are compatible with their previous lives (Loftus-Farren, 2017).

Another major issue relating to climate change relocation is how to fund it and find alternatives. As already stated, securing adequate funding is a very difficult task and there is no Federal policy nor much funding to assist those communities who will bear the burden of climate change (Kim, 2019). Tribes that are attempting to meet the challenge and relocate include the following, the Northwest tribes Quinault, Hoh, Quileute and Sauk-Suiattle and the Southeast tribes of the Isles de Jen Charles Band of Biloxi-Chitimacha-Choctow. In Alaska, 31 villages are facing imminent risk from climate change due to coastline erosion or flooding. Thus far, 3 villages Kivalina, Shishmaref and Newtok have decided to move (Loftus-Farren, 2017). Ultimately, relocation of tribal nations will become a larger issue for many of the coastal tribes in the near future.

Introducing the Institute for Tribal Environmental Professionals (ITEP)

Set against this dynamic background, we now introduce the Institute for Tribal Environmental Professionals (ITEP) at Northern Arizona University's (NAU) and its work with US tribes and climate change. We present this as another example of the diversity and agency exhibited by Indigenous tribes in the United States, as they collectively seek to address the challenges of climate change. At the request of Hopi Tribal Elders, ITEP was created in 1992. Assisted by ITEP's original director, the late Virgil Masayesva, Professor William Auberle and the US Environmental Protection Agency (USEPA), ITEP has become the largest and only Institute to provide continuous national environmental services to tribes. ITEP's vision is to achieve "*A Healthy Environment for Strong, Self-Sustaining Tribal Communities*" and meets a critical need to increase tribal capacity and strengthen tribal sovereignty across the United States. A premier tribal training organization in the US, it has directly served 95% of all 573 federally-recognised Tribes and Alaskan Native Villages.

The environmental training programs that ITEP host range from small one-person programs in the start-up phase; to large and complex monitoring and regulatory programs and encompass in-person and on-line training, technical assistance, mentoring, and policy development in climate change, air and water quality, solid and hazardous waste, emergency response, policy and environmental code development. ITEP has also pioneered an innovative and substantive Tribal Climate Change Program (TCCP).

ITEP Tribal Climate Change Program (TCCP)

In 2009, in recognition of the ongoing impacts of climate change on their tribes, and the need to actively find capacity to respond, ITEP and the USEPA developed the ITEP Tribal Climate Change Program (TCCP). This program's brief is to build tribal capacity for addressing climate change by providing climate change training, informational resources, and adaptation planning tools to tribes throughout the US (Wotkyns & Gonzalez-Maddux, 2014). The TCCP was the first program in the nation to begin the process of assisting tribes with climate change education, mitigation and adaptation. The TCCP was originally funded by the USEPA, however, it is currently primarily funded by the US Department of Interior Bureau of Indian Affairs (BIA). These partnerships have allowed for all of ITEP's climate change services to be delivered at no cost to tribal professionals.

Early TCCP Years

When the TCCP began in 2009, much of the public still did not understand climate change. The general lack of knowledge about climate change was amplified in First Nation country due to the lack of access to communication. To address this issue, the emphasis in the first 5 years of the TCCP was on educating tribal professionals about the basics of climate change. The curriculum then advanced to include more emphasis on climate change adaptation planning. During the early years of TCCP, the development and implementation of climate change plans was hindered by the lack of a community champion, funding, staff and tribal prioritisation. The sheer geographical scale of tribes is also a challenge – tribes in the US encompass almost every state in the union, with the largest number of tribes in the state of Alaska – thus reaching a diverse set of tribes in their regions is a rigorous task. To address these challenges, ITEP set up a range of management mechanisms one of which was a Tribal Climate Change Advisory Committee.

The Tribal Climate Change Advisory Committee (Advisory Committee) is composed of a panel of six members from tribes/tribal organisations, six from federal agencies, and eight tribal resilience liaison staff. They meet with the ITEP once or twice a year and maintain strong communication. The Advisory Committee's role is to assist with the development of new courses and training materials, and it oversees an active website. It offers a port of call for others seeking advice on tribal climate matters. For example, in 2019, the US Congress requested information about tribes and climate change impacts and the Advisory Committee responded with a comprehensive response about the impacts and solutions. In 2020, the Advisory Committee was a great source of guidance for input into the US' first *National Tribal CC Forum*. In assisting ITEP to in turn help tribal professionals/communities to increase their adaptation to climate change, the Advisory Committee has become a critical component to ITEP's success.

The TCCP offers four core programs: (i) training via delivery of climate change courses, (ii) a suite of communication and mapping tools, (iii) technical and tailored assistance to tribes, and (iv) support in adaptation planning. Two of these will be described here to give some insight into how the TCCP works in practice (i) the training programs and (ii) tribal climate profiles.

Training Programs

ITEP has developed a wide range of training courses that assist tribes to build climate planning. These courses provide training in how to protect traditional knowledge (TK) and often include an Elder from the tribal community who addresses and guides the participants as well as youth. The bringing together of different tribes within a region also enables a sharing of ideas about how to combat impacts they all have in common. As they are often dealing with many of the same climatic issues, they share their data, findings and best practices, so that other tribes do not have to 'reinvent the wheel' when drafting their climate change plans. *Talking Circles* are used at the start and end of each course to help participants introduce and learn from each other.

The learning program begins with an introductory course called *Climate Change 101 – Introduction to Climate Change Adaptation Planning*. To start this course TCCP staff work with local tribal experts for at least 6 months to prepare the course and ensure the curriculum meets the needs of the region and audience. Training aims to build the capacity of tribes and resource managers to engage in climate change adaptation planning. Each course usually has about 20–25 tribal environmental professional participants from a region of the US. Location-based focused curricula enable participants to establish Climate Change Plans for their own regions. For example, a course in the northwest region will focus on ocean acidification and salmon related issues, whereas, in the southwest, the instruction would focus on drought related issues. Topics such as wild fires and flooding would be considered baseline topics due to their pervasiveness in all regions. ITEP organises the courses by region, i.e. southwest, northwest, Alaska, northeast, north and south central and southeast (ITEP, 2019). Learners are guided through a five-step planning process known by the acronym **S.A.D.I.E.**: **S**cope and engage; **A**ssess Vulnerability and Risk; **D**etermine options and actions; **I**mplement and monitor; and **E**valuate and adjust. An integral part of the course is a place-based field trip, where participants are taken to tribal locations to see and experience examples of tribal climate projects.

A second course called *Topics in Climate Change Adaptation Planning and Implementation* is an on-going series of webinars that was established to serve the many tribal professionals and public community who cannot attend an in-person course. The goal of the course is to provide an easily accessible way for participants to learn about emerging information on climate change impacts, planning methods, tools, and solutions, and find out about the latest funding opportunities. The webinars provide a forum through which participants can directly ask questions of practitioners and experts in the climate change field.

Finally, ITEP have a third online course called the *201 Cohort Course*. This course provides an opportunity for tribes to advance their skills and knowledge and occurs over an extended time period of 18 months. It requires long-term commitment that involves the facilitation of cohorts of tribes through the entire process of developing an adaptation plan. The course is divided between in-person meetings, many web-based modules, and assignments that will culminate in a completed draft climate change adaptation for each tribe that participates. Participants are helped by the cohort instructors throughout and are mentored by other tribal professionals that are also completing their plans (ITEP, 2019).

Underpinning the training and planning for tribes is a strong outreach and communication strategy which provides climate change outreach and communication materials that are culturally sensitive and relevant for tribal populations. At the heart of all of ITEP's tools is the ITEP Tribal Climate Change Program Website where current and accurate information is constantly being curated on a website that receives over 20,000 hits annually. It is the base for all the TCCP activities. Since 2015, the TCCP has also been producing a free monthly Tribal Climate Change Newsletter which has an ongoing listserv of about 1850 recipients including tribal resource managers, and representatives from government agencies, universities and tribal colleges, non-tribal organisations, the general public, and international subscribers.

The ITEP Tribal Climate Change Adaptation Toolkit

Another component of ITEP's work has been the development of the Tribal Climate Change Adaptation Toolkit. The Toolkit contains a range of resources that provide background material for adaptation planning, and a checklist and templates for Tribal Climate Change Adaptation Planning. Associated with this toolkit is the ITEP Tribal Climate Change Resources 'Mind Map': another key resource that helps build agency and capacity of tribes to respond to climate change. Launched in 2018, the Tribal Climate Change Resources Mind Map is an easily accessible "one stop shop" designed to show tribes how to find resources that will assist them with their climate change capacity building and adaptation planning.

ITEP Climate Change Tribal Profiles

The ITEP Staff also develop Climate Change Tribal Profiles which are then posted on the ITEP website. The purpose of the Profiles is twofold. First, they summarise the tribal region by explaining the uniqueness of the land, water, air, species, major regional impacts, state-wide initiatives, and some cultural issues. Second, the Profiles highlight the work that tribes are currently engaging in that may not have yet been published. This tool also allows tribes to see what is currently occurring throughout the US and to potentially help or request assistance for that tribe. ITEP staff develop the Profiles with tribal staff through personal interviews. The Profiles are intended to be short one to two-page documents that summarise the tribes' current and projected climate change impacts and how the tribes are addressing those impacts through planning and actions. The Profiles serve as examples for other tribes to see how they could identify their own climate challenges, how to evaluate and then address them in their own climate change plan process. All the Profiles are approved by the tribe before they are posted on the ITEP climate change website: there are now over 60 Tribal climate change profiles on the website. An example of a few profiles is provided below, simply to show the richness of activity each tribe is enacting in various regions.

Alaska

Tlingit and Haida Indian Tribes: This profile discusses the climate change adaptation planning activities of the Tlingit and Haida Indian Tribes of Alaska. When the tribes of southeast Alaska began seeking out climate change data specific to their region, they could not find relevant data. After 16 tribes passed resolutions calling for more data, several tribes partnered with ITEP to develop a workshop investigating the impacts of climate change specific to southeast Alaska. Stemming from that workshop, the Central Council of the Tlingit and Haida Indian Tribes of Alaska drafted their first Climate Change Adaptation Plan (CAP). In addition, they created a template available to all tribes in the region who also want to author their own plan.

Central U.S.

Confederated Salish and Kootenai Tribes (Montana): While the Whitebark Pine tree has been in steady decline for decades, the Confederated Salish and Kootenai Tribes used it as a high priority species as they researched and drafted their Climate Change Strategic Plan. Valuable for preserving high altitude snowpack, the plan now plays a key role in the repopulation of the Whitebark Pine tree both on tribal lands and beyond.

The Sac and Fox Nation of Missouri in Kansas and Nebraska (Kansas and Nebraska): The Sac and Fox Nation of Missouri in Kansas and Nebraska is tapping into the strength of collaboration among tribes in the EPA Region 7 to improve communication, build resiliency, and directly address vulnerabilities in the face of climate change. Water resources and water management are essential components of the vulnerability assessments being developed by the tribes, and training sessions will aid the climate adaptation planning each tribe is undertaking.

Midwest

Bad River Band of the Lake Superior Tribe of Chippewa Indians (Wisconsin): The Bad River Band of Lake Superior Chippewa (Ojibwe) has seen unprecedented flooding events in the last several years, in keeping with predictions on changing weather patterns. These flooding events impact their infrastructure, cultural resources, and natural resources. Yet they continue to thrive as a community, due in large part to their communication and cooperation with neighbouring tribes and their own Natural Resources Department.

Menominee Indian Tribe of Wisconsin: The Menominee Tribe has long been a model of forest stewardship resulting in a profitable timber industry based on sustainable yields. In the 1980s, oak wilt, a deadly tree fungus, began appearing on their land. The tribe recently partnered with the Climate Change Response Framework to repopulate areas affected by oak wilt with tree species that are adapted to a changing climate.

1854 Treaty Authority organisation (Minnesota): The 1854 Treaty Authority is an inter-tribal resource management agency governed directly by the Bois Forte Band of Chippewa and Grand Portage Band of Lake Superior Chippewa. In support of their responsibility to preserve, protect, and enhance the treaty rights in the 1854 Ceded Territory, they successfully developed a Climate Change Vulnerability Assessment and Adaptation Plan in collaboration with the Bois Forte and Grand Portage Bands, as well as the Fond du Lac Band of Lake Superior Chippewa.

Northeast

Aroostook Band of Micmacs (Maine): Looking to promote healthy, local food options in the face of climate change and overfishing, the Aroostook Band of Micmacs has developed a successful land-based fish hatchery. Now, in conjunction with their community farm and market, the tribe can make fresh fruit, vegetables, and brook trout, available to their people.

Southwest

Bishop Paiute Tribe (**California**): The Bishop Paiute Tribe has established multiple programs to both mitigate and adapt to climate change. They have a growing Rooftop Solar Program, Food Sovereignty Program, and Conservation Open Space Area all of which complement one another.

La Jolla Band of Mission Indians (**California**): Over the past years, Southern California has seen an increase in both severe drought and devastating wildfires.

This trend is anticipated to continue with climate change. In response, the La Jolla Band of Luiseño Indians has recently drafted a Forest Management Plan that will be part of a larger, overarching Integrated Resources Management Plan. The tribe also intends to incorporate previously completed Drought and Pre-Disaster Mitigation Plans, both of which, have already received approval at the federal level.

Summary

In the United States, First Nations continue to be impacted by the legacy of trauma caused by colonisation and subsequent overlapping institutional and policy barriers that inhibit First Nation capacity to address climate change. There is a history of genocidal treatment and policies towards First Nations; every possible effort was made to eradicate tribal nations and tribal lifeways. Despite the funding now offered by the Bureau of Indian Affairs (BIA, 2019), a lack of power and persistent poverty are consistent challenges.

As this chapter has outlined, despite that grim history, tribes and tribal peoples have survived and are thriving today. Indeed, many tribes believe that the trauma they endured has prepared them for this current era of climate change. Tribes continue to tirelessly protect their original homelands and to teach their children their lifeways. Tribes are building their future through their Climate Change Plans which will protect their land, water ways and seas. They have a resilient spirit and believe that despite all the atrocities that may come, they will prevail and lead during these times. Tribes continue to teach their children their languages, stories, prayers, songs, dances, prophecies, and lifeways. One prophecy by a Lakota Holy Man in the mid nineteenth century said that it would take seven generations for the *hoop to heal.* That 7th generation walks with us today. They are the hope to heal our planet for the next seven generations.

References

Bennett, T. M. C., Maynard, N. G., Cochran, P., Gough, R., Lynn, K., Maldonado, J., Voggesser, G., Wotkyns, S., & Cozzetto, K. (2014). Indigenous peoples, lands, and resources. In J. M. Melillo, T. C. Richmond, & G. W. Yohe (Eds.), *Climate change impacts in the United States: The third national climate assessment* (pp. 297–317). U.S. Global Change Research Program.

BIA [Bureau of Indian Affairs]. (2019). *Tribal resilience program.* BIA [Bureau of Indian Affairs]. https://www.bia.gov/bia/ots/tribal-resilience-program

Browning, P., & Aegerter, G. (2019). Washington's fawn sharp elected president of national congress of American Indians. *KUOW News.* https://www.kuow.org/stories/indian-country-elects-washington-s-fawn-sharp-national-president

Burkett, M., Verchick, R. R. M., & Flores, D. (2017). *Reaching higher ground – Avenues to secure and manage new land for communities displaced by climate change.* Center for Progressive Reform. http://progressivereform.org/articles/ReachingHigherGround_1703.pdf

Crosby, A. W. (1976). Virgin soil epidemics as a factor in the aboriginal depopulation in America. *The William and Mary Quarterly, 33*(2), 289–299.

CTKW [Climate and Traditional Knowledges Workgroup]. (2014). *Guidelines for considering traditional knowledge in climate change initiatives.* https://climatetkw.wordpress.com

Dermansky, J. (2019). *Isle de Jean Charles Tribe turns down funds to relocate first United States 'climate refugees' as Louisiana buys land anyway.* Desmog – clearing the PR pollution that clouds climate science. https://www.desmogblog.com/2019/01/11/isle-de-jean-charles-tribe-turns-down-funds-relocate-climate-refugees-louisiana

DOE [Department of Energy]. (2015). *Climate change and the United States energy sector: Regional vulnerabilities and resilience solutions.* U.S. Department of Energy. https://www.energy.gov/policy/downloads/climate-change-and-us-energy-sector-regional-vulnerabilities-and-resilience

Fixico, D. L. (2018). *When native Americans were slaughtered in the name of 'civilization'.* A&E Television Networks. https://www.history.com/news/native-americans-genocide-united-states

ITEP. (2019). http://www7.nau.edu/itep/main/Home/

Jantarasami, L. C., Novak, R., Delgado, R., Marino, E., McNeeley, S., Narducci, C., Raymond-Yakoubian, J., Singletary, L., & Whyte, K. P. (2018). Tribes and indigenous peoples. In D. R. Reidmiller, C. W. Avery, D. R. Easterling, K. E. Kunkel, K. L. M. Lewis, T. K. Maycock, & B. C. Stewart (Eds.), *Impacts, risks, and adaptation in the United States: Fourth National Climate Assessment, volume II* (pp. 572–603). U.S. Global Change Research Program. https://doi.org/10.7930/NCA4.2018.CH15

Kim, G. (2019, November 2). Residents of an eroded Alaskan village are pioneering a new one, in phases. *NPR News.* https://www.npr.org/2019/11/02/774791091/residents-of-an-eroded-alaskan-village-are-pioneering-a-new-one-in-phases

Koch, A., Brierley, C., Maslin, M. M., & Lewis, S. L. (2019). Earth system impacts of the European arrival and Great Dying in the Americas after 1492. *Quaternary Science Reviews, 207*, 13–36.

Krogstad, J. M. (2014, June 13). *One-in-four native Americans and Alaska natives are living in poverty.* Pew Research Center. https://www.pewresearch.org/fact-tank/2014/06/13/1-in-4-native-americans-and-alaska-natives-are-living-in-poverty/

LII. (2019). *Section 137.10. Definitions.* Cornell Law School Legal Information Institute, Cornell University. https://www.law.cornell.edu/cfr/text/42/137.10

Loftus-Farren, Z. (2017). Losing home. *Earth Island Journal.* Berkeley, California. http://www.earthisland.org/journal/index.php/magazine/entry/losing_home/

Ma, M. (2018, November 14). *New resources support tribes in preparing for CC.* University of Washington News. https://www.washington.edu/news/2018/11/14/new-resources-support-tribes-in-preparing-for-climate-change/

NCA3. (2014). *National climate assessment: Indigenous peoples.* https://vimeo.com/100528198

NCAI. (2013). *Securing our futures report.* National Congress of American Indians (NCAI). http://www.ncai.org/resources/ncai_publications/securing-our-futures-report

NCAI Website. (2019). National Congress of American Indians Website. *About NCAI.* http://www.ncai.org/about-ncai

Norton-Smith, K., Lynn, K., Chief, K., Cozzetto, K., Donatuto, J., Redsteer, M. H., Krugar, L. E., Maldonado, J., Viles, C., & Whyte, K. P. (2016). *Climate change and indigenous peoples: A synthesis of current impacts and experiences.* United States Department of Agriculture.

Stebbins, S., & Sauter, M. B. (2019). These are the 25 worst counties to live in: Did yours make the list? *USA Today News.* https://www.usatoday.com/story/money/2019/03/13/worst-places-live-us-counties-ranked-poverty-life-expectancy/39163929/

Walker, M. (2019, October 15). Fed up with deaths, Native Americans want to run their own health care. *The New York Times*. https://www.nytimes.com/2019/10/15/us/politics/native-americans-health-care.html

Warner, K. (2015). *Tribes as innovative environmental 'laboratories'*. University of Kansas School of Law Working Paper, University of Kansas.

Wotkyns, S., & Gonzalez-Maddux, C. (2014). *CC adaptation planning, training, assistance, and resources for tribes. Report developed for US federal agencies working on tribal climate change programs and initiatives*. Institute for tribal Environmental Professional, Northern Arizona University. http://www7.nau.edu/itep/main/tcc/docs/resources/RptCCAdaptPlanningTribes_2014.pdf

Chapter 5
Ethnic Minorities, Traditional Livelihoods and Climate Change in China

Lun Yin

Introduction

Global climate change has had an increasingly serious impact on the agricultural practices of Indigenous peoples around the world, especially forms and types of agriculture, animal husbandry, hunting and gathering. Yet an interactive relationship between climate change and Indigenous traditional livelihoods persists. In this chapter, the ways in which climate change has impacted on the agricultural traditions of ethnic minorities[1] in China is explored, and the ways in which traditional knowledge is used to adapt to those changes described.

In China, there are 55 officially recognised ethnic minorities. The traditional livelihoods of ethnic minorities are diverse, and include agriculture and nomadism, as well as hunting, fishing and gathering. In the past 30 years, nomadism and hunting in China have been abandoned due to the implementation of government policies that have created herdsmen settlements and promulgated laws that prohibit hunting. As a result, ethnic minorities who used to make their living mainly by nomadism and hunting have turned to other means of survival (Pei & Zhang, 2014; Zhang, 2010; Zhengwei, 2010).

In the Qinghai-Tibetan Plateau, adjustments to grazing have occurred, including the collection of goods such as caterpillar fungus to sell to urban markets and a shift to a greater reliance on seasonal cash jobs (Fu et al., 2012; Haynes & Yang, 2013; Yin, 2011). In the Yunnan region, farmers have adapted their planting patterns and market dynamics to respond to climate driven water stress, using their traditional

---

[1] In this chapter, we refer to Indigenous peoples as ethnic minorities, as that is the cultural term used in China

---

L. Yin (✉)
Southwest Forestry University, Kunming, China
e-mail: 13888267735@163.com

© The Author(s) 2022
M. Nursey-Bray et al., *Old Ways for New Days*, SpringerBriefs in Climate
Studies, https://doi.org/10.1007/978-3-030-97826-6_5

knowledge to help them survive climate impacts (Shaoting, 2010; Su et al., 2012; Wenhui, 2010). Elsewhere in the Yunnan region, ethnic groups have established 15 local knowledge-based climate adaptation actions, partly supported by government or private sector support, which has enabled the establishment of initiatives such as water storage, drought relief and irrigation programs (Li & van Dijk, 2012).

Some, such as the Tu (see Perspective 3.4), have transformed their dependence on agriculture to embrace tourism.

Perspective 5.1
Climate, Tourism and Building Capacity for the Tu People of China's Qinghai Province
Haiying Feng and Victor Squires

The rate of temperature change on the Qinghai-Tibet Plateau is up to four times the rate in China overall (Liang et al., 2014; Xu & Liu, 2007). Alpine areas such as this are very sensitive to temperature changes, especially as glaciers begin to melt and permafrost thaws (Feng & Nursey-Bray, 2020). In particular the region has experienced 'warmer and drier' conditions, and local ecologies are changing (Feng & Nursey-Bray, 2020; Feng & Squires, 2020). As temperatures rise there has been upslope migration of some plant species (and their attendant insect, bird and rodent cohorts).

The Tu people, one of the 55 officially recognised ethnic minority groups in China, have adapted, mitigated and integrated climate changes into their cosmologies. For the *Tu*, landscapes, life and human activities are determined by the cycle of the seasons. Until recently, most Tu in the rapidly changing rural settlements practiced sedentary agriculture, supplemented by minimum animal husbandry, and seasonal work in towns and cities. This has ended abruptly in the face of climate change impacts that threaten their subsistence livelihoods which are dependent on a highly variable climate, and further complicated by the increasing frequency of extreme events such as snow disasters (that can wipe out up to 30 per cent of livestock, Liu et al., 2014), and melting glaciers that cause local flooding. However, it is in their capacity to focus on the everyday, on the micro-experiences of change, which enables them to reveal how they see, feel, and make sense of climate change in their own lives.

Specifically, the Tu have responded and adapted to the climate impacts on their livelihoods by taking advantage of growing interests in cultural heritage and cultural and eco-tourism. The government has invested huge sums to foster Public-Private-Partnerships that engage businesses to invest in accommodation and restaurants and other service facilities, and local people have been encouraged to provide bed and breakfast type accommodation and convert local heritage buildings into guest houses. In Huzhu (an autonomous Tu county in eastern Qinghai) a purpose-built cultural park has been established to preserve some historical buildings, artefacts and provide interpretive exhibitions.

Local Tu people are employed as guides, service and maintenance staff. Many find work in the service industry (eg. hotels, restaurants, transport) and others are involved in tours to scenic and/or historically-significant spots outside the main settlements. There has been a transformational change as the local Tu people have actively transitioned from small-plot farmers to entrepreneurs and service providers. This type of policy formulation and its implementation by the government in minority autonomous areas such as those inhabited by the Tu, builds capacity for ethnic minorities such as the Tu who are victims of ecological and climatic changes beyond their control.

Indigenous people like the Tu will be forced to adapt rapidly by both traditional and contemporary means, including the use of traditional knowledge, agro-pastoral innovation, and the further development of tourism-based economies. The switch from being dependent on the highly variable, and often inadequate, precipitation to produce subsistence food crops, to a higher level of income and an expansion of opportunities that come from involvement in cultural and eco-tourism, is a major adaptation to climate change (Photo 5.1).

Photo 5.1 The Huzhou people where cultural and eco -tourism is an alternative mode of adaptation. (Credit Haiying Feng)

Traditional Types of Agriculture

Ethnic minorities have a long history in agriculture and due to their diversity, a wide range of types of agriculture have evolved and been adopted (See Table 5.3). Agriculture is the sector where the greater range of climate adaptation occurs and to understand its breadth, it is important first to understand what the key types of agriculture currently exist in China.

Table 5.3 Traditional types of agriculture in China

Types	Ethnic minorities	Location (Province)	Climate condition	Altitude (m)
Shifting agriculture	Wa, Jinuo, Jingpo, Bulang, Dulong	Yunnan	Warm, humid	1200–2000
Terraced agriculture	Hani, Miao, Yao, Zhuang, She, Dong	Yunnan, Guizhou, Guangxi, Zhejiang	Warm, temperature difference changes greatly	120–3000
Paddy field agriculture	Dai, Dong	Yunnan, Guizhou, Guangxi	Sweltering, humid	80–600
Oasis agriculture	Uygur	Xinjiang	Sweltering, dry	30–1200
Agro-pastoralism	Tibetan	Yunnan, Sichuan, Tibet, Qinghai, Gansu	Cold, temperature difference changes greatly	1900–4000

Shifting Agriculture

Shifting agriculture has a long history in China, and while it used to be the main livelihood for many ethnic groups today it is now only common among the mountain ethnic groups in the Yunnan Province such as the Jingpo, Bulang, Jinuo and Dulong ethnic groups (Shaoting, 2001).

In the Yunnan Province, shifting agriculture still survives as a farming mode as the region is subtropical and tropical, with warm or hot climates. Controlled by southeast and southwest monsoon zones, it receives adequate rainfall and has abundant forest biological resources. Such geographical and climatic environments provide a good ecological basis for the local people to engage in shifting agriculture. Secondly, the groups that practice shifting agriculture are mostly found in mountainous areas, within small basins and river valleys. The mountainous terrain is complex and steep, so it is difficult to build irrigation systems and overall difficult to cultivate paddy fields in these places: shifting agriculture can thus be practiced.

Terraced Agriculture

Terraced agriculture occurs in the Yunnan, Guangxi, Guizhou, Sichuan and other provinces in Southwest China, and also in Hunan, and Zhejiang to name a few. The ethnic minorities chiefly engaged in terraced agriculture include the Hani, Miao, Yao, Zhuang, She and Dong. Among these groups, the Hani in Yunnan have the largest scale of terraced agriculture: it is also the most famous and was accorded world cultural heritage value by UNESCO.

The Hani people live in a mountainous area with an altitude of 1400–2000 m. The climate is mild and the rainfall is abundant. They live between two forests – deciduous broad-leaved forest and evergreen broad-leaved forest, both very suitable for rice planting (Qinwen, 2009). Therefore, the Hani people in Yunnan Province have formed the most centralised and largest terrace agriculture in China. For the Hani, undertaking terraced agriculture is a practice closely related to the local environment; terraces are built in a subtropical monsoonal area, but due to high altitudes valley areas are extremely hot and dry, while the mountain area has low temperatures and more precipitation.

Paddy Field Agriculture

Ethnic minorities living in basins and river valleys at low altitudes are engaged in paddy field farming, such as the Dong and Dai minorities. Most of these ethnic groups are located in Southwest China, especially in the Guangxi, Guizhou and Yunnan provinces. Over time, a large number of glutinous rice varieties have been developed in the paddy fields of ethnic minorities, enabling rich agricultural biodiversity and development of a complex traditional knowledge system.

Paddy field agriculture occurs between 100 m and 600 m above sea level, needs flat terrain, abundant rainfall, crisscrossing rivers, and hot and humid temperatures which form the environmental basis for rice planting. While planting rice, the Dong people in Guizhou also raise fish and ducks in their paddy fields. This "rice-fish-duck" symbiotic system represents a compound livelihood system based on paddy field agriculture (Haiyang, 2009). In China, this unique composite livelihood system reflects significant agricultural cultural heritage.

Oasis Agriculture

Oasis agriculture mainly occurs in the south of Xinjiang and records show they have been an ongoing practice for more than 4000 years. Oasis agriculture is dependent on local topographies and rivers: after the melting of glaciers and snow, rivers form which flow down to the desert areas. After encountering the dry and high temperatures of the desert, the water is gradually absorbed and evaporated, and finally, at the base of the mountains, the river gradually disappears and forms an 'oasis' in the desert: this is the basis of oasis agriculture. Over 2000 years ago, during the Han Dynasty, local people formed many small city states based on oasis agriculture. Today the Uygur people continue to engage in oasis agriculture in this region. This type of agriculture occurs between 30 m and 1200 m above sea level, from mountain basins through to alluvial plains. Due to the difference in altitudes, terrain and climate in various regions, different farming systems have evolved and a variety of crop species are cultivated. The main crops are wheat, barley, corn, rice, chestnut, sorghum, soybean, millet (Yun, 2010).

Agro-Pastoralism

Agro-pastoralism is the final mode used by ethnic minorities to build livelihoods; and is very important to the Tibetan people in North-West Yunnan (Eastern Himalayas), not only as a means of survival but also for biodiversity resource management and culture. For agro-pastoralists there is a mutual dependence between agriculture and herding as both agriculture and herding are important to the livelihoods of the Tibetan people providing basic and necessary food to local people, manure for crops, and crop straw as fodder to livestock. Additionally, harvest farmlands become grazing lands for herding during winter seasons (Lun & Zachary, 2018). Tibetan people mainly raise yak, cattle yak, yellow cattle and goats and they usually herd animals according to different seasons and physical locations. Herding is classified in three ways: summer grazing land (summer alpine pasture), spring-autumn interim grazing land, and winter grazing land. In the Tibetan language, local people call summer grazing land "Ru La", which means grassland covered by snow. At an altitude of around 4000 m, this area is also referred to as alpine meadows. At the same time, local people call the spring-autumn interim grazing land as "Ru Mei", which means grassland located at middle altitudes – in this case around 3000 m and referred to as meadow and sloping land. Finally, winter grazing land is known as "Ru Bo", which means grassland located near village houses. Its altitude is around 2000 m and refers to sloping land near the village. While villagers herd yak in the alpine pastures and goats in the winter grazing lands respectively, they will herd cattle yak and yellow cattle regularly amongst all three grazing lands.

Traditional Knowledge of Agriculture

Although there are many studies that investigate the impact of climate change on agriculture, including the traditional modes of agriculture used by ethnic minorities, they usually rely on Western scientific data, models and formulas to draw 'objective' and scientific conclusions. Yet the traditional knowledge of the ethnic minorities, especially their perception of climate change, has not been given due attention. Ethnic minorities in China however, are *the* practitioners of traditional agriculture. Their observation and perceptions come from real life and while they may not be 'scientific', they directly reflect the impact of climate change on local livelihoods and further contain wisdom in implementing adaptations to climate change. All the forms of agriculture described above have been built on intimate knowledge of the local environment, including climate. In fact, different climatic conditions have been one of the factors that explain agricultural farming diversity. Ethnic traditional knowledge systems encompass local knowledge about water, forest, weather and climate systems. The next section describes some elements of this knowledge for each of the agricultural modes (Photo 5.2).

Photo 5.2 Climate field school – learning about traditional knowledge for climate change. (Credit: Lun Yin)

The Phenological Calendar in Shifting Agriculture

Like all agriculture, shifting agriculture is closely related to climate, and more dependent on seasonal and climate change than others. Most of the ethnic minorities who have engaged in shifting agriculture in Southwest China have few written records of climate and phenology, and no written calendars. But they have maintained, via oral inheritance, a wealth of traditional knowledge and experience about climate and phenology, represented by their phenological calendar. A phenological calendar charts the division of seasons and defines the seasonal rhythm of shifting agriculture in a year, to form a series of farming etiquettes. For example, within this phenological calendar, the most important thing is to control the time of tree felling, burning land and sowing: together these actions prescribe the relevant farming etiquette. The production process of shifting agriculture is thus determined by the phenological calendar of the specified season, and the cycle and change within the phenological calendar marked by multiple farming etiquettes.

Customary Law of Terraced Agriculture

Water resources by contrast, are the foundation for terrace agriculture. The most remarkable feature of terraced agriculture is its capacity to make full and effective use of water resources by using the stereoscopic climate, landform and the natural environment characteristics of 'how high the mountain is, how high the water is'. Over generations, ethnic minorities have created a customary law system for water resource management, which is an important part of their traditional knowledge system. For example, the Hani people exercise very strict water resource management customary laws. These customary laws include the distribution of water resources, the protection of water forests, punishment of water theft and the maintenance of irrigation channels (Hui, 2019). Under the constraints of customary law, the ecosystem on which terraced agriculture depends is protected, especially the forest and water resources. The strength of this knowledge is demonstrated by the fact that during 2009–2012, the Yunnan province suffered a number of serious droughts, and agriculture suffered a huge loss: yet the Hani terraced agriculture was not affected due to its complex knowledge system about water management.

Technology of Paddy Field Agriculture

In paddy field agriculture, ethnic minorities have developed a fine-tuned technology of change and adjustment based on their traditional knowledge of the local climate. For example, the Dong people engage in paddy field agriculture in high altitude areas, and rely on underground well water and spring water where the water temperature is relatively low. Over the years, to improve the water temperature, the Dong people built ditches from the water source to guide the well water or spring water to the paddy fields. Via the construction of these ditches, the Dong people have developed a reticulated water flow system with the effect that long-distance circular flows ultimately improve the water temperature; when water finally flows into the paddy field it is conducive to the growth of rice (Kangzhi, 2019). Other ethnic minorities have similar traditional technologies and this long-term process of water management remains an enduring traditional technology that enables active and ongoing adaptation to the local climate and environment through the accumulation and application of generations of experience and knowledge.

Water Conservancy Project of Oasis Agriculture

Due to its desert location, the most important thing for oasis agriculture is to ensure that there are sufficient water resources for irrigation. In this context, the Uygur people invented a water conservancy technology, developed through the accumulation of generations of experience and traditional knowledge. Called 'Karez' it is a channel system divided into two parts: an underground channel and an over ground channel. The maximum length of the underground trench can reach up to 14 km, while the minimum length is about 3 km. Every 10–30 m there is a vertical well which is directly connected to the ground by the underground trench; the deepest vertical well is 60–80 m. The underground channel rises gradually, and finally as it is exposed to the ground becomes the over ground channel (Yi, 2019). According to the earliest written records, Karez has been applied for over 180 years and represents great innovation and the inventiveness of Uygur traditional knowledge. Karez is bound by a strict management system and includes its own maintenance and distribution of water resources. The Karez system is of great significance to the success of ongoing oasis agriculture, especially in areas experiencing climate related water shortages, and it guarantees the sustainable development of local agriculture.

Local Perceptions About the Impact of Climate Change to Agriculture

Despite the strong and generational corpus of traditional knowledge that ethnic minorities in China have accumulated, modern climate change presents new challenges and observation skills honed over centuries are now being used by ethnic minorities to assist in building adaptation to climate change. Contemporary perceptions of climate change moreover are multi-faceted, and not only limited to the climate field, but include the phenology of natural environment and livelihoods, that is to say, the observation of climate and phenology change forms the perception of ethnic minorities on climate change.

For example, for the Dai People rising temperatures and the accompanying drought, affects the output of some traditional rice varieties that are not resistant to high temperature. Temperature increases have affected the Tibetan animal husbandry: the early maturity of forage, the decrease of forage yield at high altitude, and the degradation of pasture, the increase in diseases and insect pests and other phenomena, make livestock breeding more challenging.

Climate instability has also made some farming calendars invalid – for the Jino people climate has changed the whole process of shifting agriculture processes and farming etiquette. At the same time, unstable precipitation also affects the growth of crops, reducing grain harvests. For the Hani people climate instability not only brings some challenges to water supply resources but has also led to an increase in crop disease and pests; sudden cooling also leads to crop freezing in high altitude areas.

For the Wa people, heavy rainfall and continuous high temperatures, cause flooding that completely destroys hillside crops. For the Uyghur, continuous drought is affecting their oasis agricultural systems, not only reducing its output, but also amplifying waste. Heavy snowfall, drought, debris flow, landslides and other disasters have caused huge losses to Tibetan Alpine agriculture.

At the same time though, climate changes have wrought some positive and beneficial effects. For example, Uyghur farmers have observed that due to climate warming and increases in precipitation, they have, in some regions been able to sow crops earlier and as autumn frosts are arriving later, the total growth periods are longer, and yields per unit area increased. Tibetan farmers have been able to grow some crops that were previously unable to survive due to the cold climate.

Traditional Knowledge and Climate Change Adaption

As climate impacts are increasingly observed and experienced, the ethnic minorities of China are also actively adapting to these impacts. Their specific responses highlight how adaptation based on local observations can provide levels of detail at scales unencumbered by global climate change models. The application of their traditional knowledge is central to this. In the context of building adaptation for agriculture, traditional knowledge is incorporated into adaptation in three ways; (i) the traditional use of agricultural bio-species and genetic resources; (ii) traditional technical innovations for bio-resource use and traditional practices for farming and living styles; and (iii) traditional cultures such as customary laws and community protocols that are related to agriculture. What does this look like in practice?

The Tibetans use of language is a good example – they draw on their linguistic appellations for seed species, which delineate how each species copes with climatic conditions at much finer detail than Western scientific ones. As such when climate change such as drought and high temperature occurs, local Tibetans will choose to breed cattle and highland barley according to the capacity of different varieties to adapt to the impact of climate change. For the Dai, Hani, Wa and Dong people, traditional knowledge which provides baseline understanding of crop resistance and survival relating to heat, drought and rainfall, is being used to determine the choice of appropriate rice varieties.

Traditional knowledge is also an innovative process. In order to adapt to climate change, ethnic minorities have adopted a series of agricultural technical innovations that specifically build on existing knowledge to respond to future climate challenges. These traditional technical innovations include: staggered seed crop planting, mixed cropping, crop rotation practices, soil fertility improvement practices and soil tillage practices.

As cultivated land is distributed across different altitudes, shade and sun regimes, dry zones and wetlands, there is great diversity amongst the climate borders areas under cultivation whether for agro-pastoral or terraced agricultural purposes. Staggered seed crop planting has been developed by the Tibetan and Hani peoples as it can be used as an adaptation to the microclimate environment of different plots, while reducing the risk of crop failure due to the instability of rainfall and drought. Such staggered seed crop planting can maximise the use of sunlight, temperature and water resources, while reducing the risk of rainfall instability.

Mixed cropping is another common way of shifting agriculture. Farmers of the Wa, Jino and other ethnic groups often plant two or three or more crops in the same farmland creating mixed plantings according to the division and the utilisation of different types of land.

At the same time, due to the change of altitude and slope, there are also differences in temperature in different areas of the same plot. There are differences in thickness, fertility and barrenness between steep and flat areas. Therefore, different crops or different varieties of the same crops are mixed in a small plot. For example, the Jino people will plant different varieties of early rice, middle rice and late rice in the same plot, and then beans, sorghum and vegetables. This practice of mixed planting across different plots enables the different varieties of upland rice and different crops to be cultivated on the plots with suitable climates and environments: yield can be increased. Mixed planting in the same plot can thus avoid the risk of ending up with no harvest as a result of an extreme weather event caused by climate change.

For the Blang people, the rotation of *gramineae* and *solanaceae* species is often adopted. Specifically, there are three forms of rotation: cotton and upland rice rotation, cotton, upland rice and corn rotation, or cotton, upland rice, coix and corn rotation. Cotton, as a heat loving crop, has the advantages of drought tolerance, has strong resistance to disease and insects, and brings other advantages such as enhancing soil and water conservation.

In addition to the use of traditional knowledge as a means by which to invent new technical solutions, traditional agricultural knowledge also incorporates customary law systems and protocols for the use of water resources as well as watershed management. This includes the use of methods to prepare for anticipated hazards and to reduce the risk of climate change. Ethnic minorities use two forms of customary law: one that exists between different ethnic groups, and another that exists within the same ethnic group.

The Yunnan region is a good example: here the residential pattern of mountain ethnic minorities is based on different altitudes. From 100 m to 1000 m above sea level, the river valley and basin areas is mainly inhabited by the Dai people who engage in paddy field agriculture. From 1000 m to 2000 m above sea level, the semi mountainous areas are inhabited by the Hani people who engage in terraced field agriculture. From 1500 m above sea level, the mountainous areas are mainly inhabited by the Wa, Jinuo and Bulang people who engage in shifting agriculture and finally, the plateau area (at an altitude of more than 2000 m) is mainly inhabited by Tibetans who are engaged in agro-pastoralism.

When different ethnic groups live in the upper and lower reaches of the same small watershed, during droughts, the different groups will formulate common customary laws to reasonably and fairly distribute agricultural irrigation water, thus avoiding potential conflicts that might be caused by water shortages. However, when residents of a small watershed live in the same ethnic minority, such as Tibetans, customary laws will also be formed within the same ethnic group to distribute water resources reasonably. In any one village, there will be a collective meeting of villagers held inside the village to discuss how to allocate water resources. The Tibetan village will be divided into two sections according to their geographical position. In this case, customary law determines that farmers who live in the upper part of the village can irrigate their farmlands in the morning and farmers living in the lower part of the village can do so in the afternoon. For several villages, village meetings are also required to arrange the use of water resources, so that the upstream and downstream villages can take turns, thus achieving fair use of water resources. In this way, the direct and ongoing application of ethnic customary laws about water resource sharing, avoids contemporary conflicts over water use during drought and creates efficiencies in the use of water resources, enhancing the resilience of ethnic minorities to adapt to climate change.

Conclusion

Climate change is affecting Indigenous peoples around the world, including ethnic minorities in China. In the process of adapting to climate change, the traditional culture, technology and knowledge of ethnic minorities in China has been shown to play an active role in adapting to it. Traditional knowledge can improve understanding of the influence of climate change on livelihoods, and also offer ideas around how to develop locally driven and appropriate ways to adapt to climate change. The use of traditional knowledge to create adaptation responses, ensures that they are appropriate to the geographical location and environment of ethnic regions, and overall more effective.

References

Feng, H. Y., & Nursey-Bray, M. (2020). Adaptation by herders on the Qinghai-Tibetan Plateau in response to climate change, and policy reforms: the implications for carbon sequestration and livelihoods. In Z. Shang, A. A. Degen, M. K. Rafiq, & V. R. Squires (Eds.), *Carbon management for promoting local livelihood in the Hindu Kush Himalayan (HKH) region*. Springer.

Feng, H. Y., & Squires, V. (2020). Socio-environmental dynamics of Alpine Grasslands, steppes and meadows of the Qinghai–Tibetan Plateau, China: A Commentary. *Applied Sciences, 10*(18), 6488.

Fu, Y., Grumbine, R. E., Wilkes, A., Wang, Y., Xu, J., & Yang, Y. (2012). Climate change adaptation among Tibetan pastoralists: challenges in enhancing local adaptation through policy support. *Environmental Management, 50*, 607–621.

Haiyang, C. (2009). Dong culture and the maintenance of traditional livelihood. *Journal of Guangxi University for Nationalities, 5*, 27–35.

Haynes, M., & Yang, Y. (2013). Adapting to change: transitions in traditional range management of Tibetan yak herders in northwest Yunnan. *Environment Development and Sustainability, 15*, 1065–1077.

Hui, Z. (2019). Terraced farming culture in western mountainous areas. In Y. Shaoting (Ed.), *The general aspiration of national culture in western China: agricultural roll* (pp. 123–173). Yunnan People Press.

Kangzhi, L. (2019). Paddy farming culture in the Western lowlands. In Y. Shaoting (Ed.), *The general aspiration of national culture in western China: agricultural roll* (pp. 177–270). Yunnan People Press.

Li, H., & van Dijk, M. (2012). Climate change and farmers responses in rural China. *International Journal of Water, 6*, 290–310.

Liang, Y., Ganjurjav, W. N., Zhang, B., Qing-zhu, G., Luo-bu, D., Zhuo-ma, X., & Yu-zhen, B. (2014). A review on effect of climate change on grassland ecosystems in China. *Journal of Agricultural Science and Technology, 16*, 1–8.

Liu, F., Mao, X., Zhang, Y., Chen, Q., Liu, P., & Zhao, Z. (2014). Risk analysis of snow disaster in the pastoral areas of the Qinghai-Tibet Plateau. *Journal Geographical Science, 24*, 411–426. https://doi.org/10.1007/s11442-014-1097-z

Lun, Y., & Zachary, M. (2018). Traditional knowledge and community-based research to support climate change adaptation in eastern Himalayas. In S. J. P. Tume & V. I. Tanyanyiwa (Eds.), *Climate change perception and changing agents in Africa & South Asia* (pp. 159–170). Vernon Press.

Pei, Q., & Zhang, D. D. (2014). Long-term relationship between climate change and nomadic migration in historical China. *Ecology and Society, 19*(2), 68. https://doi.org/10.5751/ES-06528-190268

Qinwen, M. (2009). Characteristics and protection of Agricultural Cultural Heritage of Hani Terrace. *Academic Exploration, 3*(2009), 31–37.

Shaoting, Y. (2001). *People and forests: Yunnan Swidden agriculture in human-ecological perspective*. Yunnan Education Publishing House.

Shaoting, Y. (2010). *Records of shifting agriculture in Yunnan [M]*. Yunnan People Press.

Su, Y., Xu, J., Wilkes, A., Lu, J., Li, Q., Fu, Y., Ma, X., & Grumbine, R. E. (2012). Coping with climate-induced water stress through time and space in the mountains of southwest China. *Regional Environmental Change, 12*, 855–866.

Wenhui, Y. (2010). Climate, resources and beliefs: Bai people's traditional knowledge and climate change. In Y. Shaoting (Ed.), *Chinese culture and environment* (pp. 199–206). Yunnan People Press.

Xu, W., & Liu, X. (2007). Response of vegetation in the Qinghai-Tibetan Plateau to global warming. *Chinese Geographical Science, 17*, 151–159.

Yi, L. (2019). Oasis farming culture in the South of Tianshan. In Y. Shaoting (Ed.), *The general aspiration of national culture in western China: Agricultural roll* (pp. 275–341). Yunnan People Press.

Yin, L. (2011). Local perspective on climate change of Tibetan people – case of Guonian village in Deqin County, Yunnan. *Think Front, 37*(4), 24–28.

Yun, Q. (2010). *Silk road: Oasis study.* Xinjiang People Publishing House.

Zhang, W. (2010). Types and early warning mechanism of Tujia proverbs. In Y. Shaoting (Ed.), *Chinese culture and environment.* Yunnan People Press.

Zhengwei, G. (2010). The adaptation of Dong people to climate change in Hubei Province from the perspective of farming. In Y. Shaoting (Ed.), *Chinese culture and environment.* Yunnan People Press.

Chapter 6
Do Not Forget the Dreaming: Communicating Climate Change and Adaptation, Insights from Australia

Melissa Nursey-Bray, Robert Palmer, and Phil Rist

Introduction

In this book so far, we have described a diversity of adaptation responses and touched on some of the challenges inherent in building Indigenous led adaptation, including a lack of resources, loss of traditional knowledge and the way in which colonisation and globalisation has caused ongoing trauma such that climate change further amplifies existing issues.

In this chapter, and using Australia's Aboriginal and Torres Strait Islander peoples as a focus, we extend these considerations to reflect on the cross-cultural interfaces, particularly communications, that influence the success and implementation of Indigenous driven climate change adaptation activities on their Country.[1] The chapter draws from a 2-year project, funded by Australia's National Climate Change Adaptation Research Facility (NCCARF) that investigated Indigenous priorities for climate change adaptation. The study included a survey, two workshops and semi-structured interviews with Aboriginal groups from South Australia, New South

---

[1] In Australia, the term 'Country' is used to denote the land and seas to which a person is traditionally affiliated, it is their nation, and one is intrinsically connected to and part of it. See Bird Rose (1996) for more detail

M. Nursey-Bray
Department of Geography, Environmentand Population, University of Adelaide, Adelaide, SA, Australia
e-mail: Melissa.Nursey-Bray@adelaide.edu.au

R. Palmer
University of Adelaide, Adelaide, SA, Australia
e-mail: robert.palmer@utas.edu.au

P. Rist
Girringun Aboriginal Corporation, North Queensland Land Council Girringun Aboriginal Corporation, Cardwell, QLD, Australia
e-mail: eo@girringun.com.au

© The Author(s) 2022
M. Nursey-Bray et al., *Old Ways for New Days*, SpringerBriefs in Climate Studies, https://doi.org/10.1007/978-3-030-97826-6_6

91

Wales, Victoria, Queensland, Northern Territory and traditional owners from the Torres Strait. The research helped identify some Indigenous priorities for managing and adapting to climate change impacts, but specifically, it drew our attention to the significance and role played by climate communications and the ways in which they could be tailored, and then used to encourage Indigenous people to participate in adaptation actions, as well as build cross cultural knowledge collaborations.

Climate Challenges

Aboriginal and Torres Strait Islander peoples in Australia are not just thinking about future and projected change, they are experiencing it in the here and now (Bird et al., 2013; Nursey-Bray et al., 2013, 2019; O'Neill et al., 2012; Petheram et al., 2010; Zander et al., 2013; DCCEEW, 2022). Climate change science shows that in Australia, there will be higher temperatures, ocean warming, hotter, and earlier and more frequent bushfires over extended periods, changes to fruiting and flowering regimes, disruptions to reproductive cycles in wildlife and fewer cyclones that will yet be more intense when they do occur. These climate changes will have multiple implications, one of which is the impact of sea level rise, which may result in potential dispossession and threaten cultural survival.

In Tasmania, coastal sites subject to sea level rise and wind erosion have affected significant rock art sites: in Preminghana, a significant Aboriginal rock art site on the West Coast, based on very friable crumbly rock, is under threat (TAC, 2012). Rock engravings are now commonly submerged by the sea and are protected only by seaweed. In another beach in southwest Tasmania, old established trees have been undercut by water action as a result of recent changes in the high-water regime. Along this eroding coastline there are five rock art sites as well as shell middens which are being destroyed by the rising sea. Middens are occupation sites, where Indigenous peoples over millennia have met and eaten together – the middens are often metres deep accretions of shells that represent what was harvested over the centuries. These midden sites are central to the identity of Tasmanian Aboriginal peoples, especially women who, as Cockerill (2018, 1) notes:

> will barely recognize their submarine landscapes, and they will feel this ripple through their cultural identity. Living midden sites will be washed away… the process of cultural recovery among Aboriginal Tasmanians will be further hindered by weather, waves, species and other hallmarks of the Anthropocene.

As Emma Lee describes (Perspective 6.1), these impacts affect gendered cultural identity as well.

Perspective 6.1
Identity in a Time of Rising Tides
Emma Lee

I have been an invited subject in both Jessica Cockerill's evocative essay, *Hauntology on country* (Cockerill, 2018), and Jen Evan's critical think-piece, *Giving voice to the sacred black female body in Takayna country* (Evans, 2018), to describe how Indigenous bodies are the geography of place in Tasmania, Australia. These writings highlight how, in particular, women are physically embodied by the connections to country. By this, our bodies are constructed in tandem with the things of country that make meaning in our lives. In Tasmania, sea country is where women's governance power resides, where we make of ourselves the roles, responsibilities and obligations to family and country through inhabiting and reflecting marine resources.

Identity, then, as a *trawlwulwuy* woman of *tebrakunna* country is bounded by the relationship with sea country. I have always had a connection with the living midden sites of our country – the physical stories of our diets, lives and knowledges – told in the heaped shells, tools, ochre, shelters and human burials that rim and hug the coastline of Tasmania. To be part of a midden and to have my body reflect the contours and histories of these astonishing places is to define myself as an Indigenous woman.

Rising seas are not unknown for a culture that is 40,000 years old in Tasmania; our island home was once connected to the mainland of Australia. The land bridge was flooded more than 10,000 years ago and changed our histories. The midden culture had to be built anew, as our living floors of the past were subsumed by the great tides. Adaptation and resilience to change were previously not insurmountable challenges, we were able to rebuild our geography and create new middens that have lasted since then. As women, our identities and connections to sea country were not torn, but were a consistency and harmony that repaired and balanced the damage of those colossal sea changes.

The comfort I have in knowing that our women and their bodies were a marker for cohesion in creating the midden cultural heritage, giving rise to my identity, is tempered by the fact that those conditions are unlikely to be repeated today. Colonisation has wrought trauma and disruption to our lives and stolen our rights to continue to make our coastal geography as women of the sea. No more can we manage sea country, through the representations of middens, by providing food for our families, giving birth and being buried in them, singing the songs and applying the ochre that give life to our bodies and place.

The added pressure of the immediate dangers we live in today such as additional sea level rises and warming waters, has also meant that the middens that defined my body and identity are being annihilated. Middens I grew up with are now gone; the tides have washed them away. This gives rise to a question: am I still a *trawlwulwuy* woman if our middens are gone? How can I adapt my body when my connectors to sea country are destroyed? Do I drown alongside my cultural heritage?

These are not easy questions to face. Nor does it seem the responses are easily identifiable or accessible. The characteristic of my heart belongs to these places, they make me who and what I am today. Yet I must believe that there is an answer somewhere, that my very being will not disappear. Perhaps I am able to adapt to the immediacy of identity and middens and use memory and oral histories as the gateway to connection, but somehow this leaves me with unfulfilled desires to *belong*.

In my dreams I imagine that a great, future revolution will occur that allows us to return to our ways and rebuild our middens, in the process redefining our connections to new spaces, species and agency. However, hope is a poor substitute for action. I do not know the answers, but there is a glimmer for me as just one individual woman from *tebrakunna* country.

We survived then – through the rushing waters and again later when the strangers came - and we will survive now.

Somehow, I need to shift the tangible identity markers of middens into a statement of belonging in other ways. Perhaps this is what resilience means, to never give the power to define my identity to someone or something else.

I do not blame sea country for taking back what we made of her, the middens that represent our connections. Sea country has her own agency, her own story and journey. Resilience and adaptation may mean that I forge a personal peace with the life-giver that is each wave encroaching on my identity. Perhaps this is the lesson and story, that she remains and I go, regardless of my thoughts, identity and connections. Sea country is eternal, she will teach future generations the lessons that need to be known and these will be different from the lessons that I have embodied.

Adaptation to identity is personal and myriad yet connected with a life far greater than just this *trawlwulwuy* body. Sea country will guide me, guide many more to come, and the skills to read her, listen and know, do not necessarily change – just the answers. Identity, then, may well be nested not in the things that I once thought, but the ability to keep those old ways refreshed, invigorated and alive (Photo 6.1).

Photo 6.1 Tasmanian Sea Country. (Credit: Emma Lee)

Emma's story finds resonance with the Ngarindjeri people of the Coorong in South Australia, where middens and sites of cultural significance are also at threat of erosion and sea level rise (Wiltshire, 2019).

However, many other kinds of cultural sites will be at risk. Coastal Aboriginal quarries across Australia (and related stone tool making workshops, often containing thousands if not millions of artefacts), are now often found out of place due to flooding. These quarries and associated workshops are rare and critically important cultural sites to Aboriginal peoples and their destruction due to climate change will cause irreversible cultural damage to their peoples.

Uluru, in central Australia is another example. Predicted to experience even hotter weather, drought, fire and flash floods, these factors will combine to affect the morphology of the rock in such a way that it may 'weather' the rock so that it will cause caverns, that will lead to a honeycomb effect, literally changing the face of Uluru. This impact will have profound implications for the Anangu people for whom Uluru is a sacred site (Hughes et al., 2018) (Photo 6.2).

Photo 6.2 Uluru, a famous Anangu site, will be affected by climate change. (Credit: Melissa Nursey-Bray)

In the areas hardest hit, forced or involuntary migration may be needed, and those in remote or rural areas will be forced to relocate or migrate. This is a serious issue in the Torres Strait, (the island region at the very top of Australia), where the displacement of islanders and thus migration to mainland Australia is likely

(Hennessy et al., 2007). Yet, even when faced with the knowledge of climate change and the need to relocate, Elders in the Torres Strait simply do not want to move:

> When the high tide and strong winds come together, it breaks. We pray we don't lose our homes. We don't want to leave this place. (Dennis Gibumam, cited in Roache, 2019, 1)

Security of tenure remains fundamental to achieving appropriate policy and cultural responses to climate change in Australia.

Many culturally important traditional harvest species for Australia's Aboriginal and Torres Strait Islander peoples, such as turtle and dugong are also going to be impacted by climate change. For example, the Green Turtle has a reproductive cycle that is heat sensitive, with males produced in cooler temperatures and females in hotter temperatures. Over time, due to increased heat, more females than males will hatch causing reproductive imbalance and ultimately affecting hunting regimes. Similarly, as sea level rise and erosion cause some beaches to wash away, turtles – who have a magnetic imprint that enables them to return to the same beach to breed – will (and are) find it hard to return to the same beach to breed (Brothers & Lohmann, 2015). Increased storms, ocean acidification and warming will also affect dugong (another key cultural species), and the seagrass habitats on which they depend, affecting their range and distribution over time, and in turn cultural activities associated with them (Cavallo et al., 2015).

Many Australian Indigenous groups have also observed the impact of climate change on 'bush tucker' species, which now flower at different times, or not at all. In many cases, increased heat, or changes to precipitation and water regimes have restricted the distribution of key and important species. In turn these changes are affecting traditional knowledge as these species are the 'signs' by which people read the landscape and which are embedded within millennia old knowledge systems, are now changing (Photo 6.3).

Photo 6.3 Wildflowers such as these in Arabana country are scarcer due to fewer rainfall events. (Credit: Melissa Nursey-Bray)

> Bush flowers – it's make us sad, things are not the same, we used to get out and we used to
> get so many flowers, so many kinds, now you just get a few and there not whole areas in
> flower, used to pick whole bunches! (Arabana respondent, cited in Nursey-Bray et al., 2020)

In tropical savannas, such as Kakadu National Park, (in northern Australia), which is the country of the Bininj/Mungguy people, climate change impacts will affect sources of sustenance such as water birds, fish, turtles, magpie geese, crocodiles and freshwater food plants such as waterlilies (Bowman et al., 2010; Ibbett, 2010; Leger & Fisk, 2016).

It is not just natural and cultural systems and practices that are being impacted by climate change. Climate policy frames consistently overlook Indigenous realities in Australia, including the fact that the majority of Indigenous peoples live in cities. While 75% of Aboriginal and Torres Strait Islanders live in cities, 75% of funding goes to those living on their Country, or are invested in land and sea management programs. Cities today thus remain the locations that most embody the continued colonist occupation of traditional territory, or as Cockerill (2018, 1) states so eloquently, "Western colonisation is a haunting that started with genocide and continues with the Anthropocene". This situation creates an erasure of Indigenous peoples in decision making and policy about how to address climate impacts in both urban and remote contexts (Nursey-Bray et al., 2022).

It also means that socio-economic impacts are amplified by climate change. Along with many other Indigenous peoples who live in urban centres, including in Australia, New Zealand, Canada, the United States, Bolivia, Brazil, Chile, Venezuela, Norway and Kenya (consistent with a broader trend towards global urbanisation), many urban Aboriginal and Torres Strait Islanders live in poor suburbs, in cheap, poorly built housing that are not well insulated or protected against heat, cold, or flooding (Brand et al., 2016; Horne et al., 2013). Often multiple people live in the same house, again making adapting to climate change very difficult. Heat stress caused by higher temperatures is harder to adapt to when many people live in insufficient and unsustainable housing, and further many cannot afford nor have access to air conditioning (or air conditioning that works) (Horne et al., 2013). In town camps in Alice Springs, Australia, heat stress and other impacts cause hospitalisation or death (Low Choy et al., 2013). Diseases such as dengue fever, are also predicted to become more widespread; people that live in tropical regions will face accelerated climate related disease risks (Russell et al., 2009).

Adapting to Change

Despite these challenges, Indigenous Australians have been active in pursuing adaptation and mitigation for their country and people. Some peoples such as the Arabana, the Torres Strait and Yorta Yorta have been developing their own adaptation strategies. For example, the Torres Strait Climate Change Strategy 2010–2013 asserts that:

> This Torres Strait Climate Change Strategy provides our region with a guiding framework and action plan so we can proactively address current and potential impacts by identifying the range of priority responses required based on sound scientific research and community involvement. (Torres Strait Regional Authority (TSRA), 2014, 1)

Other groups focus on the documentation of climate change, development of indicators and seasonal calendars, (often led by Elders) as modes of knowledge transmission. The Miriwoong people in the Northern Territory have produced an interactive seasonal calendar, showing ongoing and future weather patterns, so as to help ensure younger generations to adapt to those changes. Some groups, including Girringun Aboriginal Corporation and the Northern Australia Indigenous Land and Sea Management Authority (NAILSMA) focus on the building of networks and partnerships with Western scientists to build on Country programs, including wider management models such as co-management. Other groups have harnessed their TEK about fire to work with governments and researchers to build carbon farming programs.

The West Arnhem Land Fire Abatement Project is a case in point; it is a partnership between the Aboriginal Traditional Owners and Indigenous ranger groups of the plateau, Darwin Liquefied Natural Gas (DLNG), the Northern Territory Government and the Northern Land Council. Through this partnership, Indigenous ranger groups are implementing strategic fire management across 28,000 km^2 of Western Arnhem Land to reduce the frequency of wildfires and so offset some of the greenhouse gas emissions from the Liquefied Natural Gas plant at Wickham Point in the city of Darwin.

Other groups incorporate adaptation planning within wider management plans. For example, the Ngarrindjeri people explicitly address climate change in their Sea Country plan (Ngarrindjeri Tendi (2006), called Yarluwar-Ruwe:

> We recognise the huge impacts of global warming on our lands and waters and all living things. In recent years we have observed changes in our local environment that tells us that climate change is a reality. We see that the breeding behaviour of birds is changing, and the fruiting and flowering of our bush foods is changing. We have watched our fresh water holes dry up or turn salty and we've seen our coastal camping places and middens washed away by rising sea levels. When we lose these places we lose not only part of our cultural heritage, but we also lose an irreplaceable record of Ngarrindjeri adaptation to climate change in the past. (Ngarrindjeri Tendi, 2006, 19)

At a community level, different groups tailor their responses to local circumstances. In Kownayama, Queensland, adaptation work focusses on sustaining wetlands, fire monitoring, town and weather planning and feral weed and plant eradication. Others like the Alinytjara Wilurara in South Australia, have developed climate 'report cards' which provide an assessment of climate impacts, so they can work out how to progress adaptation.

Currently there are also over 2100 Indigenous rangers who work within Ranger Centres across Australia to look after their traditional Countries. Many of these Ranger groups are now focusing on how to address and adapt to climate change. One example is the outreach activity run by the Dhimurru and Yirrkala Rangers in the Northern Territory. These Rangers based in North-East Arnhem Land work with their local schools to deliver a *Learning on Country* program about understanding climate change and its effects on seasons and the abundance of natural resources. In this program, Elders take students to important places in their region to teach/share cultural stories about important sites, sea-levels, tides, seasons and changes to the landscape through song and dance.

However, despite this array of adaptation innovations which show how Australia's Indigenous peoples are employing both traditional and innovative approaches to cope with impacts of climate change and variability, finding 'fit' for them is still a challenge. There remain issues in achieving effective collaboration at both state and federal levels between Indigenous peoples and the government agencies responsible for climate change policy planning and development. Further, there continues to be challenges in the dialogue between western scientific researchers and Indigenous peoples as they seek to work out how to respectfully utilise traditional knowledge for adaptation planning. For example, for the Anangu people in South Australia, despite their extensive ecological knowledge - there remains a need to develop mechanisms that allow traditional ecological knowledge to co-exist with scientific knowledge to assist future planning and management of natural resource systems (Bardsley & Wiseman, 2012).

We argue that these tensions between Indigenous peoples and policy makers are amplified due to systemic inappropriate communication and engagement. Top-down institutional processes still dominate policy in ways that do not facilitate Indigenous voices and do not adequately recognise traditional cultures and practices. In July 2021, a National First Peoples Gathering on Climate Change, affirmed that Indigenous peoples "want to set their own agenda on climate knowledge and action". In reflecting on these limitations, we argue that appropriate communication tools will help find fit for Indigenous initiatives within broader governance regimes and acknowledge the diversity in adaptation responses currently being brokered by Aboriginal and Torres Strait Islanders across Australia.

Bespoke Climate Communications Are Key for Effective Indigenous Climate Change Adaptation

For the remainder of this chapter, the focus will be on exploring how the communication interface between Australia's Indigenous peoples and those who are working in the climate change and adaption field may be progressed. Working with Indigenous groups in Australia (when there are so many) has particular challenges and we do not pretend that our findings represented in this chapter are representative, they are insights only that may provide guidance relating to future communications.

The first step is to seek out and learn the Indigenous communications landscape. Langton et al. (2012) argue that specific communications in an Australian Indigenous context are needed to help build policy responses at an institutional level. Choy et al. (2013) view Indigenous climate change communications as a necessary support mechanism for the development of collaborations between different stakeholders, noting that Indigenous communications and engagement should be designed to "ensure that the next generation of Aboriginal communities are across climate change adaptation to address issues of succession planning" (Choy et al., 2013, 2). For Green and Minchin (2014), communications and engagement are mechanisms used to help reduce health impacts in Indigenous contexts associated with climate change.

Climate change adaptation communications are also a tool for spreading knowledge about climate change itself (Moser, 2014). Appropriate communications can spur Indigenous peoples into action, and encourage their active involvement in projects that are designed to help Aboriginal and Torres Strait Islander peoples adapt to climate change (Nursey-Bray & Palmer, 2018). For Arbon and Rigney (2014, 482), communications and engagement are "the heartbeats - the centrering, the conveying of knowledge between non-Indigenous and Indigenous participants" and designed to help build relationships and resolve conflict (Cochran et al., 2008). Appropriate communications will also encourage Indigenous community participation in climate initiatives.

Different Understandings of Dominant Terminology Have Implications for Communications

The challenge of working out how to present climate terminology so it is culturally appropriate and has resonance is a key first task. As discussed in Chap. 1, key terms such as 'vulnerability', 'resilience' and 'adaptation', either do not resonate within different Indigenous groups or mean something completely different. Gaps in understanding between Indigenous and Western knowledge systems and in turn management are thus created. The different ways in which vulnerability and resilience paradigms are understood by Indigenous groups and science are illustrative of this dissonance: there is a systematic dichotomous conceptualisation in the literature (and within governance regimes), of Indigenous groups as either 'vulnerable' and/or 'resilient'. This automatic adherence to the view of Indigenous peoples as vulnerable and or resilient can in turn, entrench existing and historical (i.e., colonially derived) structures that make the 'other' invisible in decision making, additionally complicating the way climate change knowledge is communicated (Nursey-Bray et al., 2020). Cameron (2012, 4) reflects that the very way scientists talk about vulnerability can limit the ways in which Indigenous peoples can have an input: "buttressing political and intellectual formations that underwrite a new round of dispossession and accumulation in the region". Or as Maru et al. (2014) observe, Indigenous peoples often in fact have a dual narrative of vulnerability *and* resilience.

In Australia, our collaborations revealed a rejection of tropes around vulnerability and resilience in favour of assertions of agency and survival. The Arabana people of the Kathi Thanda-Lake Eyre region for example, conceived of themselves as neither vulnerable nor resilient. Instead, they saw themselves as *survivors,* whether of climate change, or colonisation:

> We survived colonisation, we are fragmented and damaged but we survived; all that happened in such a small space of time, how is that different to, well, it is exactly the same with climate change, and we will adapt to that too, again maybe not unscathed, but we will survive. (Arbon, cited in Nursey-Bray et al., 2020)

> We not weak! – 'vulnerable' you say?! We still here. Always have been, always will be…look, when *you* come to take our land away, when *you* come talk to us about climate, well, *you* might gotten burned in the sun (!) but *we'll still be here.* We're a strong people, not goin anywhere… (Marree respondent 4, cited in Nursey-Bray et al., 2020)

> Country looks after us, we look after country, we survive together. We adapt together. Over generations and generations. (Port Augusta Respondent 5, cited in Nursey-Bray et al., 2020)

Significantly, we found that the term 'adaptation' is also understood in different ways: Western science explains adaptation as a natural autonomous process (what species do to respond to change) but also asserts it in the literature as a way of understanding how society responds to climate impacts. However, Indigenous peoples propose that adaptation is something that has *always* happened and is rooted in socio-political and cultural contexts. This view asserts that climate adaptation is not an abstract (or new) idea connected just to *climate change,* but one rooted in history and time connected to country and *everything* relating to it.

Adaptation then is not a new practice or concept, but a practice with ancient origins and framed in cultural/historical terms:

> Well that idea you have of adaptation, you fellas a bit Johnny come lately with that one! We been adapting to change for ever. You might try take our language, our land, our customs, but we survived all that. Now maybe you want to help us adapt to that? Adaptation?, well that good to help get us jobs, get us some rangers to look after this country we got back now, that'd be good. (Marree Respondent 7, cited in Nursey-Bray et al., 2020)

In this sense, adaptation is the application of *old ways to new days* and in being constructed as such generates a sense of confidence and a positive outlook that while it is an issue that needs to be overcome, it *can* nevertheless *be* done. However, these terms are understood, or represented, we suggest that making the attempt to understand the different ways in which different Indigenous groups understand terms like climate, change, vulnerability, resilience and adaptation is a critical and necessary first step in the embedding and co-creation of appropriate forms of climate change and adaptation communications.

Ensure Messaging is Culturally Aligned and Connected to Country

As noted above, a crucial first step in communicating climate change and adaptation, is to understand and articulate how different terms such as adaptation are understood and the replacement of the dominant trope of Indigenous peoples being vulnerable/resilient with an Indigenous driven one of agency and survival is an important shift in narrative emphasis. Another question is whether or not climate communications are to be situated within positive or negative frames. In what ways do climate related communications need to be culturally attenuated; and, in a country like Australia where there are literally hundreds of (and very diverse) Indigenous groups, how is this balance sought? Who, moreover, will be the carriers, the messengers of these communications? Indigenous peoples, others, or both?

Framing theory helps to answer these questions as frames define problems, assist in the diagnosis of causes, enable evaluation of solutions and prescribe solutions that make communications "more noticeable, meaningful or memorable" (Entman, 1993, 53). Framing theory has been used in other research, where it has been helpful in understanding what triggers public interest in climate change news stories (Foust & O'Shannon Murphy, 2009) or to concentrate audience attention by anchoring it within a resonant cultural or social context (Lück et al., 2016). However, the use of framing theory does not tell people *what* to think, but instead presents solutions for *how* they could think about a particular issue.

Many different frames have been used to help communicators give meaning to the climate change story. Early work in this area pitched climate frames as a global issue and helped communicators target public audiences with the intent of making them understand that climate change is a global issue requiring international action based on strong scientific evidence (Palmer, 2013; Palmer et al., 2017). However, researchers argued that this approach was too abstract a notion for most people to grapple with (Painter, 2013), so other frames were introduced. For example, one frame represented climate change as an almost certain apocalypse, and a tipping point for all life on Earth (Russill & Nyssa, 2009). In 2006 climate change was predominantly framed as an absolute catastrophe threatening all life on Earth. Public polling at that time showed 68% of Australians believed "global warming was a serious and pressing problem... [and a] top-rated threat to Australia's vital interest" (Cook, 2006, 4). At that time, negative framing was also driving high-level political support to tackle climate change, with the then Australian Prime Minister Kevin Rudd describing climate change as the "great moral, environmental and economic challenge of our age" (Rudd, 2007).

However, this negative framing was challenged in favour of a positive climate change narrative where it was framed as an opportunity to develop a bright new clean and green sustainable future (Futerra, 2010). Consequently, dominant frames about climate change and adaptation shifted from a negative to a positive. Evidence suggests that this change in framing was a mistake: in Australia, as climate change communications became friendlier and positive, public polling showed that people wanting governments to take aggressive action on climate change fell from 68% in 2006 to 36% in 2013 (Hanson, 2013). The Federal election of 2022 comes full circle, with people again overwhelmingly voting for action on climate change.

This history has relevance when deciding how to communicate climate change to Indigenous peoples. Efforts by the Arabana people, the Indigenous traditional owners of Kati Thanda – Lake Eyre region of Australia, highlight the saliency of whether or not to develop positive or negative frames in Indigenous focussed climate communications. When beginning the development of adaptation for their country, it was clear that there was very little interest when positive images and frames were used to describe the climate impacts in the region. However, when the framing was then revised, to construct climate change as a negative and direct threat to the Arabana and their Country, interest was engaged and sustained at a high level for the rest of the project. In this case, the Arabana did not need to be offered saccharine versions of climate change, where impacts are softened, nor required a positive spin to be put on the information. The use of negative frames effectively 'hooked' their attention (Nursey-Bray & Palmer, 2018).

Moreover, we found that across Australia, Indigenous peoples constantly asserted their own adaptive capacity, and their confidence in the agency and resilience of their peoples despite agreeing that climate change was a risk. From their perspective, when located amongst the issues they have *already* had to adapt to, including invasion, massacre, forced relocation from Country and the effects of the Stolen Generation, climate change was constructed as simply another issue that their people will contend with. It is important to develop direct and honest communications about climate change that align with Indigenous world views and their knowledge of country.

Indigenous Humour as a Climate Change Adaptation Education Tool

In collaboration with Indigenous colleagues, we identified that the frame of engagement and how it is situated within cultural world views about climate, we found that humour could be used as a deliberate tool to communicate information in meaningful ways. Humour is undeniably a central feature of Australian Indigenous culture (Duncan, 2014). It plays a pivotal role in shaping daily lives (French, 2014), and has always been a levelling force within Indigenous communities, used as a way of exchanging knowledge and the maintenance of Aboriginal identity (Duncan, 2014). Humour in this case is a complex institutionalised practice central to Aboriginal culture and is used to "regulate social behaviour by joking and shaming tactics" (Duncan, 2014, 2). Stanner (1982, 40) notes "the underlying philosophy of Aboriginal humour is likely to baffle a European mind", because it is used in a way to deal with elements of their culture alien to a contemporary, non-Indigenous Australian. In particular, Indigenous peoples use humour to help them contend with the ongoing horror of colonisation, an experience non-Indigenous people can ever truly comprehend. Duncan (2014, 82) adds: "humour is the only way we get through hard times".

There are some examples that support the notion of using humour to frame communications about a serious issue such as climate change to Indigenous Australians. For example, Redmond (2008, 257) references a comedy performed in the Kimberly region of Western Australia, where information about white invasion is performed to an Aboriginal audience who at times were said to experience "a radical loss of bodily composure" from laughing so hard whilst watching the show. Torres Strait Islanders equally use laughter in their culture, but using humour has to be utilised with more caution, especially if the focus is on a family group or an individual's reputation who "may perceive a slight where none is intended" (Beckett, 2008, 295). The Yolngu dance troupe Chooky Dancers from Elchoe Island used humour as the basis for their dance routines to communicate their own messages about Aboriginal culture. The use of humour in climate change and adaptation communications has also been considered by Walker (2014, 368), who says it might be better "to take a playful and innovative approach in order to engage readers' hearts and not just their heads" about climate change.

Many non-Indigenous people would find it difficult to comprehend such an approach. As Aboriginal Fulbright scholar Angelina Hurley notes, Indigenous lives and cultures are still being misunderstood and appropriated (Hurley, 2017). Therefore, there is a chance that an added benefit of utilising humour in climate change communications is that Western communicators (involved in the climate change adaptation field) might be enabled to reconstruct their understanding of Indigenous peoples, thus contributing to a wider appreciation and understanding of Australia's Indigenous peoples. As McCullough (2014, 678) observes "humour creates a space wherein Murri people can talk to each other and fight against non-Indigenous understandings and perceptions of Murri life".

Social Media Is a Preferred Means of Communication

We suggest that social media is also utilised as an appropriate means of communication: its *modus operandi* aligns with ancient cultural traditions of Australia's Indigenous peoples. Indigenous peoples have always been oral and visual societies (Carlson, 2017). For millennia they adorned walls with art and used storytelling for the transferal of cultural norms and Indigenous law. While these traditions persist and are in use today, Australia's Aboriginal and Torres Strait Islanders have also moved online; they are prolific social media users and use it as a mechanism for sharing stories about important things that relate to a cultural group. Indigenous Australians use Facebook and YouTube as fundamental tools to express their voice and to help affirm cultural identity (Carlson, 2017; Rice et al., 2016). Described as a new frontier, the Indigenisation of social media in Australia means that not only do Indigenous peoples communicate with each other, but it opens up new ways of enabling others from outside of their world to connect and understand them in more meaningful ways (Carlson, 2017), including connections with their Elders, and to facilitate the transferral of important cultural information between generations (Carlson & Frazer, 2018).

Do Not Forget the Dreaming

However, even when achieving the right pitch in communications, there is an ongo-ing dilemma between knowledge and practice that has the potential to create an impasse in the development of climate communications with some Indigenous groups. For some Indigenous groups knowledge about, and articulation of, 'the Dreaming' (an English word used loosely to describe Indigenous creation narra-tives and religious belief systems), explains away the origins and causes of climate change, and potentially belief in the need and capacity to adapt to its impacts. For example, for the Gunggandji People climate change is central to their origin stories:

> Land and sea country as an integrated cultural landscape is reinforced by our creation sto-ries and Song lines that bind land and sea environments together. Often reflecting the great changes that took place when the sea rose up to flood our coastal plains of thousands of years ago. (Gunggandji Aboriginal Corporation, 2013, 10)

For some of the Gunggandji people climate change is not a product of human inter-vention, but is rather a communication from mother (Earth) that is meant to express her disgust with the way people were treating her. Thus the location of climate change as part of ancient cultural storylines, in many cases appears to engender a sense of fatalism or resignation; a sense that the situation was out of Indigenous control, power, and influence. Yet adapting to it is still conceived as possible when couched within ancestral lived history. An excerpt from the Gunggandji Land and Sea Country Plan (2013, 21) demonstrates this belief:

> Climate change is another grave challenge to our country that is not of our making. However, like all coastal groups around Australia, Gunggandji people have demonstrated the capacity to adapt to climate change over thousands of years. Our ancestors have lived through a 10-metre rise in sea level, great changes in rainfall, the arrival of new plant and animal species and the great upheavals caused by volcanic activity as river courses changed and new land forms emerged.

While this assertion of Indigenous world views must be respected, it potentially cre-ates a communication barrier. The adherence to traditional knowledge systems and a belief in adaptive capacity over time contests scientific knowledge about climate change, and its likely impacts on Indigenous peoples and their country. The act then of asserting scientific knowledge as the imprimatur for climate action could not only be perceived as an act of contemporary colonisation but simply rejected because it does not align with Indigenous world views around the causes of climate change.

Conclusion

In this chapter we have presented an overview of some of the key trends and per-spectives of Australian Indigenous peoples as they face climate change. A key emphasis in the chapter has been the central role played by communications in building Indigenous adaptation to climate change, and the need to ensure their alignment with the particular cultural needs within each cultural group in Indigenous Australia.

Indigenous peoples across Australia are extremely active in the promotion and development of climate mitigation and adaptation programs. They have done so both independently and in conjunction with policy makers and researchers. They are not afraid of climate change and draw on ancient adaptation skills and practices to manage future changes. However, climate change and dealing with it will be a joint endeavour, and this has specific consequences for the development of key narratives, messages and modes of communication that will be culturally appropriate. Humour is one mechanism that could be used. Ultimately, the challenge of communicating climate change, always a complex challenge, has specific nuances when engaging with Indigenous peoples in Australia.

References

Arbon, V., & Rigney, L. (2014). Indigenous at the heart. *Alter Native: An International Journal of Indigenous Peoples, 10*(5), 478–492.

Bardsley, D. K., & Wiseman, N. D. (2012). Climate change vulnerability and social development for remote indigenous communities of South Australia. *Global Environmental Change, 22*(3), 713–723.

Beckett, J. (2008). Laughing with, laughing at, among Torres Strait islanders. *Anthropological Forum, 18*(3), 295–302.

Bird, D., Govan, J., Murphy, H., Harwood, S., Haynes, K., Carson, D., Russell, S., King, D., Wensing, E., Tsakissiris, N., & Larkin, S. (2013). *Future change in ancient worlds: Indigenous adaptation in northern Australia.* National Climate Change Adaptation Research Facility.

Bowman, D. M. J. S., Prior, L. D., & De Little, S. C. (2010). Retreating melaleuca swamp forests in Kakadu National Park: Evidence of synergistic effects of climate change and past feral buffalo impacts. *Austral Ecology, 35*(8), 898–905.

Brand, E., Bond, C., & Shannon, C. (2016). *Indigenous in the city: Urban indigenous populations in local and global contexts.* The University of Queensland. https://poche.centre.uq.edu.au/files/609/Indigenous-in-the-city%281%29.pdf

Brothers, R., & Lohmann, K. (2015). Evidence for geomagnetic imprinting and magnetic navigation in the Natal homing of sea turtles. *Current Biology, 25*(3), 392–396.

Cameron, E. S. (2012). Securing indigenous politics: A critique of the vulnerability and adaptation approach to the human dimensions of climate change in the Canadian Arctic. *Global Environmental Change, 22*(1), 103–114.

Carlson, B. (2017). *Why are Indigenous people such avid users of social media?* https://www.theguardian.com/commentisfree/2017/apr/27/why-are-indigenous-people-such-avid-users-of-social-media. Accessed 2 Feb 2020.

Carlson, B., & Frazer, R. (2018). *Social media mob: Being indigenous online.* Macquarie University.

Cavallo, C., Dempster, T., Kearney, M. R., Kelly, E., Booth, D., Hadden, K. M., & Jessop, T. S. (2015). Predicting climate warming effects on green turtle hatchling viability and dispersal performance. *Functional Ecology, 29*(6), 768–778.

Choy, D., Clarke, P., Jones, D., Serrao-Neumann, S., Hales, R., & Koschade, O. (2013). *Understanding coastal urban and peri-urban Indigenous people's vulnerability and adaptive capacity to climate change.* National Climate Change Adaptation Research Facility.

Cochran, P., Marshall, C. A., Garcia-Downing, C., Kendall, E., Cook, D., McCubbin, L., & Gover, R. (2008). Indigenous ways of knowing: Implications for participatory research and community. *American Journal of Public Health, 98*(1), 22–27.

Cockerill, J. (2018). Hauntology on Country, *Overland*, 22 March 2018. https://overland.org. au/2018/03/hauntology-on-country/. Accessed 22 Aug 2020.

Cook, I. (2006). *The Lowy institute poll 2006: Australia, Indonesia and the world*. The Lowy Institute for International Policy.

DCCEEW. (2022). *State of Environment Report, Australia, 2022*. Australian Government, Canberra, Australia.

Duncan, P. (2014). *The role of Aboriginal humour in cultural survival and resistance*. School of English, Media Studies and Art History, Doctor of Philosophy Thesis, The University of Queensland.

Entman, R. (1993). Framing: Toward clarification of a fractured paradigm. *Journal of Communication, 43*(4), 51–58.

Evans, J. (2018). Giving voice to the sacred black female body in *takayna* country. In J. Lijeblad & B. Verschuuren (Eds.), *Indigenous perspectives on sacred natural sites culture, governance and conservation*. Routledge. ISBN 9780815377023.

Foust, C., & O'Shannon Murphy, W. (2009). Revealing and reframing apocalyptic tragedy in global warming discourse. *Environmental Communication, 3*(2), 151–167.

French, L. (2014). David Gulpilil, Aboriginal humour and Australian cinema. *Studies in Australasian Cinema, 8*(1), 34–43.

Futerra. (2010). *Sell the sizzle, not the sausage*. Futerra Sustainability Communications.

Green, D., & Minchin, L. (2014). Living on climate-changed country: Indigenous health, well-being and climate change in remote Australian communities. *EcoHealth, 11*(2), 263–272.

Gunggandji PBC Aboriginal Corporation. (2013). *Gunggandji land and sea country plan*. Gunggandji PBC Aboriginal Corporation Queensland.

Hanson, F. (2013). *The lowy institute poll 2012: Public opinion and foreign policy*. Lowy Institute for International Policy.

Hennessy, K., Fitzharris, B., Bates, B., Harvey, N., Howden, M., Hughes, L., Salinger, J., & Warrick, R. (2007). Australia and New Zealand. Climate change 2007: Impacts, adaptation and vulnerability. In M. Parry, O. Canziani, J. Palutikof, C. Hanson, & P. Van der Linden (Eds.), *Contribution of working group II to the fourth assessment: Report of the intergovernmental panel on climate change*. Cambridge University Press.

Horne, R., Martel, A., Arcari, P., Foster, D., & McCormack, A. (2013). *Living change: Adaptive housing responses to climate change in the town camps of Alice Springs*. National Climate Change Adaptation Research Facility.

Hughes, L., Stock, P., Brailsford, L., & Alexander, D. (2018). *Icons at risk: Climate change threatening Australia tourism*. Climate Council.

Hurley, A. (2017). Indigenous cultural appropriation: what not to do, *The Conversation*, November 29, 2017.

Ibbett, M. (2010). Workshop summaries: priority issues for management, knowledge gaps and ways forward. In Winderlich, S. (Ed.). *Kakadu National Park Landscape Symposia Series 2007–2009. Symposium 4: Climate Change*, 6–7 August 2008, Gagudju Crocodile Holiday Inn Kakadu National Park. Internal Report 567, January 2010, Supervising Scientist, Darwin, 111–120.

Langton, M., Parsons, M., Leonard, S., Auty, K., Bell, D., Burgess, P., Edwards, S., Howitt, R., Jackson, S., McGrath, V., & Morrison, J. (2012). *National climate change adaptation research plan for Indigenous communities*. National Climate Change Adaptation Research Facility.

Leger, L., & Fisk, G. (2016). *Kakadu – Vulnerability to climate change impacts. Case study for coast adapt*. National Climate Change Adaptation Research Facility.

Low Choy, D., Clarke, P., Jones, D., Serrao-Neumann, S., Hales, R., & Koschade, O. (2013). *Aboriginal reconnections: Understanding coastal urban and peri-urban Indigenous people's vulnerability and adaptive capacity to climate change*. National Climate Change Adaptation Research Facility.

Lück, J., Wozniak, A., & Wessler, H. (2016). Networks of coproduction: How journalists and environmental NGOs create common interpretations of the UN climate change conferences. *International Journal of Press/Politics, 21*(1), 25–47.

Maru, Y. T., Smith, M. S., Sparrow, A., Pinho, P. F., & Dube, O. P. (2014). A linked vulnerability and resilience framework for adaptation pathways in remote disadvantaged communities. *Global Environmental Change, 28*, 337–350.

McCullough, M. (2014). The gender of the joke: Intimacy and marginality in Murri humour. *Ethnos, 79*(5), 677–698.

Moser, S. (2014). Communicating adaptation to climate change: The art and science of public engagement when climate change comes home. *Wiley Interdisciplinary Reviews: Climate Change, 5*, 337–358.

Ngarrindjeri Tendi. (2006). *Ngarrindjeri Nation Yarluwar-Ruwe Plan: caring for Ngarrindjeri sea country and culture*. Ngarrindjeri Heritage Committee, Ngarrindjeri Native Title Management Committee. Accessed 5 Jan 2018.

Nursey-Bray, M., & Palmer, R. (2018). Country, climate change adaptation and colonisation: Insights from an indigenous adaptation planning process, Australia. *Heliyon, 4*(3), e00565.

Nursey-Bray, M., Fergie, D., Arbon, V., Rigney, L., Palmer, R., Tibby, J., Harvey, N., & Hackworth, L. (2013). *Community based adaptation to climate change: The Arabana*. National Climate Change Adaptation Research Facility. http://www.nccarf.edu.au/publications/community-based-adaptation-arabana

Nursey-Bray, M., Palmer, R., Smith, T., & Rist, P. (2019). Old ways for new days: Australian indigenous peoples and climate change. *Local Environment, 24*(5), 473–486.

Nursey-Bray, M., Palmer, R., Stuart, A., Arbon, V., & Rigney, L. E. (2020). Scale, colonization and adapting to climate change: Insights from the Arabana people, South Australia. *Geoforum, 114*, 138–150.

Nursey-Bray, M., Parsons, M., Gienger, A. (2022). *Urban nullius?* Urban Indigenous People and Climate Change, in Sustainability, 2022.

O'Neill, C., Green, D., & Lui, W. (2012). How to make climate change research relevant for indigenous communities in Torres Strait, Australia. *Local Environment, 17*(10), 1104–1120.

Painter, J. (2013). *Climate change in the media: Reporting risk and uncertainty*. Reuters Institute for the Study of Journalism.

Palmer, R. (2013). *Adapting communication conventions: communicating climate change adaptation to Aboriginal people*. Paper presented at Climate Adaptation 2013 knowledge + partnerships Sydney.

Palmer, R., Bowd, K., & Griffiths, M. (2017). Media preferences, low tuts and seasonal adjustment: Communication climate change adaptation to vulnerable, low socio-economic groups in Adelaide. *Global Media Journal*, Australian edition, *11*(2), 3276-1–3276-15.

Petheram, L., Zander, K. K., Campbell, B. M., High, C., & Stacey, N. (2010). 'Strange changes': Indigenous perspectives of climate change and adaptation in NE Arnhem Land (Australia). *Global Environmental Change, 20*(4), 681–692.

Redmond, A. (2008). Captain Cook meets general Macarthur in the Northern Kimberley: Humour and ritual in an Indigenous Australian life-world. *Anthropological Forum, 18*(3), 255–270.

Rice, E., Haynes, E., Royce, P., & Thompson, S. (2016). Social media and digital technology use among Indigenous young people in Australia: A literature review. *International Journal for Equity in Health, 15*, 1–16.

Roache, M. (2019, April 22). The Mayor Fighting to save he island home from climate change, *TIME*. https://time.com/5572445/torres-strait-islands-climate-change/. Accessed 1 Jan 2021.

Rudd, K. (2007). *Address to UN Bali conference on climate change*. https://australianpolitics.com/2007/12/12/rudd-address-to-bali-climate-change-conference.html. Accessed 1 Feb 2020.

Russell, R., Currie, B., Lindsay, J., Ritchie, S., & Whelan, P. (2009). Dengue and climate change in Australia: Predictions for the future should incorporate knowledge from the past. *Medical Journal of Australia, 190*(5), 265–268.

Russill, C., & Nyssa, Z. (2009). The tipping point trend in climate change communication. *Global Environmental Change, 19*(3), 336–344.

Stanner, W. (1982). Aboriginal humour. *Aboriginal History, 6*(1), 39–48.

TAC (Tasmanian Aboriginal Centre). (2012). *Preminghana healthy country plan 2015*. TAC (Tasmanian Aboriginal Centre).

TSRA. (2014, July). *Torres Strait Climate Change Strategy 2014–2018*. Report prepared by the Land and Sea Management Unit, Torres Strait Regional Authority, 36p.

Walker, L. (2014). Polar bears and evil scientists: Romance, comedy and climate change. *Australasian Journal of Popular Culture, 3*(3), 363–374.

Wiltshire, K. (2019, January 17). In the land of Storm Boy, the cultural heritage of the Coorong is under threat. *The Conversation*.

Zander, K. K., Petheram, L., & Garnett, S. T. (2013). Stay or leave? Potential climate change adaptation strategies among aboriginal people in coastal communities in northern Australia. *Natural Hazards, 67*(2), 591–609.

Chapter 7
Old Ways for New Days

Introduction

In this book, via three case studies and the inclusion of additional shared perspectives, we have explored the nature and refrains within Indigenous adaptation to climate change. While we remind the reader that we are not offering prescriptions about this issue, nor have we been able to give an example from every region of the world, these case studies highlight some commonalties of experience. Globally, Indigenous territories cover 24% of lands worldwide and contain 80% of the world's biodiversity (Etchart, 2017). This means Indigenous peoples are placed to play a significant role in addressing and being affected by big issues like climate change. However, as we have explored in this book, a range of factors need to be considered to support Indigenous adaptation and voices.

Colonisation and Adaptation

Firstly, colonisation in many Indigenous contexts remains an active agent in the formulation of adaptation strategies and as such, adaptation and adaptive governance need to acknowledge its legacy. As shown in Chap. 4, in relation to the experience of First Nations peoples in the United States, this colonial presence in many Indigenous lives today is also important when considering the enjoining of Indigenous peoples to the decision-making tables about climate change across the world. Too often Indigenous peoples are absent from the negotiating tables that decide resource use (which will still profoundly affect Indigenous groups), at much larger scales. In the Pacific and other regions, the establishment of missions and Christianity is a related but enduring colonial impact that has had substantive effects on traditional knowledge, via its imposition of prohibitions to speak Indigenous languages and practices (Nursey-Bray et al., 2021). Thus, "what may appear to be

M. Nursey-Bray et al., *Old Ways for New Days*, SpringerBriefs in Climate Studies, https://doi.org/10.1007/978-3-030-97826-6_7

politically neutral routines and procedures are important sites of contemporary colonial power, through which Indigenous resistance is managed and diffused" (Schreiber, 2006, 20).

Indigenous adaptation will be less successful if universal assumptions about Indigenous vulnerability, do not recognise the ongoing legacy of colonisation. We cannot overlook Indigenous capacity to care for their land and seas: we must all listen and engage in conversations that create ways of "seeing with both eyes, while not being blind to the hazards of colonisation" (Veland et al., 2013, 314).

Adaptation to climate change in Indigenous contexts is not just about climate change – it brings with it cultural and economic responsibilities to redress other and deep wounds and ongoing traumas. Adaptive responses then, need to be understood and engaged with in relation to what Howitt (2020) characterises the 'messy' contexts of lived experience in settler-colonial societies, where policy, science and practice "all need to develop a much more sophisticated literacy in the scale politics of response to the risk landscapes that Indigenous groups negotiate" (Howitt, 2020, 2).

Joining the Dots: Multiple Impacts Are Linked

Our case studies have also shown that any discussion around climate change also relates to the importance of livelihoods to Indigenous peoples. As the examples from the Wa and Jinhuo peoples show, amongst many others, in order to build capacity to adapt to climate change, building the capacity to maintain livelihoods is crucial. In relation to this, Maximiliano Garcia, a *makuna* of the Pirá- Paraná River notes:

> the world sees natural resources as a source of money; this is what we suffer from these days. It is not just climate change; it is changes in the way of thinking. The sacred places are a part of us, whereas the world sees these sites as a source of monetary resources to extract gold, wood. That is why climate change is abrupt, if we do not have our sacred places there is no life, for it is there that there is air, food, cures. Today it starts to rain when it is not supposed to rain, it is hot out of season, and this is what causes poverty. Because there is lack of food when it is very dry and very full, there are no crops, there are no places to hunt or fish. Nature regulates itself, there will be no fertility in the land if it rains a lot. (Nakashima et al., 2018)

Of significance here is the way in which Indigenous narratives coalesce these observations into a demand for adaptation to be built around *all* these effects, the socio-economic risks and impacts, not just climate *per se*. As the examples from Aotearoa and Sweden show, extractive violence and other appropriation of Indigenous resources run the danger of being compounded and amplified by climate impacts, and this needs to be monitored. Changes to seasons, species and ecosystems will have correlative impacts on subsistence livelihoods.

Many studies identify the holistic nature and interconnectedness of factors and risks outside of climate ones in Indigenous contexts (Tengo et al., 2014). They are not "simply 'local' but often articulate a connectedness that insists on holding global systems of economic, environmental and political governance accountable" (Howitt, 2020, 7). Thus, the way that climate impacts are linked to other activities, (as in the case of the USA and Australia) highlights that ongoing climate governance and management approaches need to acknowledge that climate change is but one of multiple issues and drivers of change affecting Indigenous peoples worldwide, and thus cannot be simply isolated in its management (Wildcat, 2013).

The Importance of Knowledge

Of course, as our case study of ethnic minorities in China highlights, not all Indigenous people live within colonial contexts, but our collaboration shows that all Indigenous peoples share a pre-occupation with, and attachment to, their traditional knowledge systems. Indigenous knowledge is an integral part of both old and new ways of adapting to climate change. Indigenous traditional knowledge has power, contributing to a rich understanding of place:

> particular formations in particular places – embodied dwelling in nature – a lived and creative relationship with the natural world (Johnson & Murton, 2007, 127).

There is growing evidence of the importance of traditional knowledge in responding to climate change (Nakashima et al., 2018; Berkes et al., 2000; Berkes, 2009; Gyampoh et al., 2009, IPCC, 2014, 864–7). The IPCC report recognises that

> Indigenous, local and traditional knowledge systems and practices, including Indigenous peoples' holistic view of community and environment, are a major resource for adapting to climate change (IPCC, 2014, 19).

The United Nations University Traditional Knowledge Initiative (UNU- TKI) has also identified over 400 cases of Indigenous peoples' being active in climate change monitoring, adaptation and mitigation, including a variety of successful strategies. The United Nations Framework Convention on Climate Change (UNFCCC) has recognised the importance of traditional knowledge, in decisions on 'enhanced action on adaptation' and within the Cancun Adaptation Framework.

Our case studies highlight the many active and present instances of applied traditional knowledge in the context of climate change. As noted by Barber (2018) in a study of the Yolngu people in Blue Mud Bay in the Northern Territory, the knowledge that the Yolngu people possess can also give them surety about their capacity to negotiate possible futures, and he cites a key reflection from an Elder confident about his capacity to adapt:

> Yolngu have been here for 50,000 years and we have survived many changes in the past. It is going to affect you guys, not me. Because I've done it in the past. If the store runs out of food, that will simply make people go back to the bush and start eating healthy again (Barber, 2018, 107).

In Norway traditional knowledge is the foundation for adaptation and the building of socio-ecological resilience to rapid change for over 200,000 reindeer and about 3000 Sami people working in reindeer husbandry (Mathiesen et al., 2018). In Southern Malawi, the integration of Indigenous Knowledge systems into adaptation provides spatial at scale information that is relevant to local adaptation to the impacts of climate change (Nkomwa et al., 2014). There are multiple opportunities for and examples of scientific and Indigenous knowledges working together, including dealing with uncertainty (Fernández-Llamazares et al., 2017). The following perspective from Australia offers another insight into the role Indigenous knowledge can play to address climate impacts, but cautions that it needs proper investment.

Perspective 7.1
Indigenous Communities Managing Wildfire
Douglas Bardsley and Annette Bardsley

There is growing evidence that a major early impact of global climate change will be experienced through changes in wildfire (bushfire) regimes (Bardsley et al., 2018). Many temperate, alpine and boreal regions are experiencing fires that are unprecedented in modern histories. Regions on the warmer and drier edge of temperate zones in California, Chile, south-eastern Australia and the Iberian Peninsula, have experienced particularly severe disasters in recent years. These changes are being caused, in part, by a warming, drying climate in association with more storm events, which together suggest that a global ecological transition is partly being driven through the agency of fire.

Indigenous traditional ecological knowledge (TEK) could be fundamental to the development of solutions to these new levels of risk. Indigenous TEK and actions offer a large range of relatively untapped opportunities for improvements in the management of all of the drivers of the wildfire hazard, while also assisting societies to follow transitional pathways to sustainable futures (Bardsley, 2018). One aspect of TEK – fire burning has had transformational impacts on landscapes and been foundational for many socio-ecosystems across the globe: relatively small numbers of Indigenous people living within North America for example had transformational impacts that drove "abrupt, climate-independent fire regime changes" (Roos et al., 2018, 8147). Anthropogenic fire was also used to manage southern European Alpine landscapes for over 7000 years (Tinner et al., 1999; Tinner et al., 2005). From 1400 BCE, it is clear that burning had opened up densely forested landscapes to facilitate hunting and agriculture (Conedera et al., 2004). Importantly, records suggest that during the period when the Roman Empire dominated, anthropogenic burning was constrained and it was not until the Middle Ages that the Indigenous Celtic knowledge could again be used to transform the forested landscapes (Conedera et al., 2007, 2017).

The exploration of, and support for, the normalisation of complex Indigenous knowledge to inform wildfire and land management would not only improve hazard and land management, but can also offer financial and socio-cultural benefits for the Indigenous people who retain, re-generate and share that knowledge. There has, for example, been a recent step-change in interest in the roles of Indigenous knowledge in burning regimes and land management in Australia.

In work we undertook with local communities in the Anangu Pitjantjatjara Yankunytjatjara (APY) Lands in north-west South Australia, we found that the wildfire risk to communities and cultural assets had increased due to a combination of climate change, higher fuel loads (less controlled burning, more invasive Buffel grass), and a lack of capacity to manage fire locally (Wiseman & Bardsley, 2016). Traditional patch burning remains a key element of land management, but Buffel grass in particular has changed the nature of fires on the APY lands, making them far more dangerous to people, settlements and cultural assets. While the Indigenous Anangu people are burning the Buffel grass in an attempt to reduce fuel loads, the fire often advantages the exotic grass because it burns very hot, damaging native species while surviving and re-shooting itself. There is thus an understandable reluctance by some people to continue to burn landscapes, especially as local traditional knowledge and cooperation to support effective burns is seen to be in decline (Bardsley & Wiseman, 2012). This finding parallels conclusions from South America by Mistry et al. (2016, 4), who note that "the current status of traditional fire management within Indigenous communities can be associated with inter-related issues of a general loss of knowledge, a breakdown of social cohesion within communities, and conflicts (particularly ideological) between Indigenous and non-Indigenous stakeholders."

Nonetheless, unique fire strategies are required for the specific places and systems, and the way to do that is to focus on people. For example, in Australia, the confidence to advocate for and undertake successful prescribed burns based on TEK requires strong cooperative approaches between traditional owners, rangers and trainees, government officers and other stakeholders to undertake burns as community events. Confidence could also be re-developed by trialling techniques within relatively low-risk situations where conditions involve low fuel-loads, high relative humidity, low-wind speeds, and large barriers to spread, such as cleared areas with relatively low ecological and cultural value. If institutional support could continue to be provided, it may be possible for Indigenous communities to generate mutual knowledge on how to manage the wildfire hazard. It would also help to "Reset the relationship" (Tebrakunna Country & Lee, 2019), between governments, societies and private landholders and Indigenous communities through the generation of mutual knowledge and action (Photo 7.1).

Photo 7.1 Patch burn in the Anangu Pitjantjatjara Yankunytjatjara (APY) Lands. (Credit: Douglas Bardsley)

Traditional knowledge is important, but we also argue that the effective incorporation of Indigenous perspectives requires an embracing of all forms of knowledge, notably *Indigenous Knowledge* (IK) in all its manifestations. Indigenous knowledge is not just the form and types of information that Western policy makers may find culturally palatable but embraces *all* forms of knowledge that is understood as knowledge and relevant by the respective Indigenous group. For example, in Australia, historical knowledge, gathered *after/as a result of* colonisation, is used by Indigenous groups to inform climate adaptation, and has become part of their cultural domain (Nursey-Bray et al., 2020). In so doing, cultural dynamism and knowledge maintenance and revival is exercised.

This process also builds on the reality that many cultures have now experienced some loss of traditional knowledge, and hence an inclusive understanding of IK *per se* provides the space for its evolution since fracture, whether derived from colonisation or globalisation. In Malekula Island, Vanuatu for instance, religious and other shifting values, require policy makers to design nuanced responses to the ongoing loss of cultural knowledge and practice:

> Understanding the historically embedded social support systems that encourage self-help and working together is likely to enhance the outcomes of donor programs and development initiatives. (Fletcher et al., 2013, 8)

In accepting all forms of Indigenous knowledge, cross generational dissemination of climate perspectives (from Elders to young people) occurs. This process acknowledges that part of the challenge of dealing with climate change today is a generational one and needs to address "the fragmentation of the culture, countryside and language" so prevalent in many communities (Sanchez-Cortes & Chavero, 2011, 386), and the importance of recognising and combining multiple forms of knowledge (Race et al., 2016).

Terminology

We also assert that language matters in developing climate adaptation collaborations with or led by Indigenous peoples. As described, in the development of ongoing effective adaptive practice Indigenous peoples have also asserted their own ways of constructing and understanding climate change and adaptation. For a start, as we have shown, the very way in which Indigeneity itself is understood varies wildly. As highlighted by the perspective on how to understand the word in African nations, even the term "Indigenous" is contested. In addition, the scientific underpinning of much climate change terminology both imposes colonial Western knowledge as ascendant, but also thereby inherently dismisses other ways of talking about climate, weather, change, adaptation, ecology and connection to place.

We argue that the automatic adherence to the view of Indigenous peoples as 'vulnerable' and/or 'resilient' can in turn, entrench existing and historical (i.e. colonially derived) structures that make the 'other' invisible in decision making, additionally complicating the way climate change knowledge is communicated. This picks up on some work that argues more generally that resilience theories need to be more critical (Weichselgartner & Kelman, 2015), otherwise they can depoliticise and hide significant structural and political issues, and create exclusionary practices by the withdrawal of support for groups deemed to be 'not resilient enough' (Porter & Davoudi, 2012). A discourse that focusses on a frame of 'a lack of' in relation to Indigenous peoples also has the effect of removing opportunities that could be taken to build agency and assert Indigenous knowledge and power into wider policy domains.

Further the entrenching of Western notions of Indigeneity and Indigenous 'vulnerability' or 'resilience', creates a 'forgetting' or blindness to the agency that Indigenous people actually continually exercise in building their adaptation programs. As highlighted in many examples throughout this book but particularly with our specific case study chapters, Indigenous agency in responding to climate change is not just a practice but a *state of mind*. As Kuruppu and Liverman (2011) highlight in a study of Kiribati, it is important to understand not just the capacity to adapt, but one's belief in one's own subjective capacity to act.

Finding 'Fit'

The way in which language and knowledge is used to reinforce dominant narratives about Indigenous peoples in climate contexts (such as the linguistic entrenching of Western norms and perspectives into climate policy) has another powerful impact: spaces are not found within governance structures to incorporate Indigenous interests and voices in climate policy, especially in urban contexts. State policy frames that do not recognise Indigenous rights or knowledge or their capacity to contribute to social and environmental recovery, both compound and contribute to how Indigenous peoples experience climate impacts and "entrench patterns of racialized disadvantage and marginalization" (Howitt & Klaus, 2012, 47).

Indigenous people's knowledge, while being increasingly valued, is seen as a resource to be *used,* as part of wider policy regimes. Indigenous knowledge is often 'harvested' to fill the gaps in current science but is not treated equally and that compromises Indigenous knowledge sovereignty (Cameron, 2012; Reo et al., 2019; Veland et al., 2013; Whyte, 2017). While the deployment of Indigenous knowledge into environmental policy arenas may be necessary, we must also consider how to support Indigenous knowledge sovereignty, intellectual property and maintenance in those same processes of policy and decision-making (Bardsley et al., 2018; Nursey-Bray et al., 2015, 2020; Nursey-Bray & Palmer, 2018). Indigenous peoples do not wish to just give their knowledge away, and this is a factor additionally complicated by the fact Indigenous knowledge is itself under pressure (Cockerill, 2018; Nursey-Bray et al., 2020; Whyte, 2020).

However, there are very few mechanisms available to ensure that there is Indigenous representation on the management committees or governance arrangements that oversee climate and adaptation policy, nor mechanisms that may assist Indigenous people to use their own knowledge to help their own communities. Further, the traditional governance modes that different Indigenous cultures use to mediate their own decisions, are not recognized at all in wider governance regimes. Important work by Robinson and Raven (2020, 2017) highlights the centrality of recognising customary law to help prevent bio-piracy and connect State and Indigenous law in relation to Indigenous knowledge and species management. Engaging with Indigenous knowledge in the Anthropocene offers insights for stewardship (Hill et al., 2020).

This also has implications for power sharing arrangements within adaptation planning. While Indigenous peoples will often seek to build partnerships with other external agencies who have the expertise to assist them implement climate change adaptation, doing so in a way that ensures they do not lose power and can maintain their own knowledge domains is an additional challenge. In Australia, Hunt et al. (2008) highlight the complexities involved in this endeavour, reflecting that: "It is simply impossible to understand the governance of Australian Indigenous communities and organisations as separate from the encapsulating governance environment of the Australian state" (Hunt et al., 2008, 3). Finding a space for Indigenous peoples within climate adaptation planning then is also about power sharing, which in turn will require a shift in management understanding of what 'local' and 'cultural 'knowledge is and the role it plays in the decision-making processes.

How can this be done? Existing scholarship has focused on how science becomes policy, how this is done, and the strengths and weaknesses of the process (Matuk et al., 2020; Diver, 2017; Jasanoff, 2004). Studies on knowledge *integration* articulate how Indigenous knowledge could be included at the knowledge interface (Hill et al., 2020). But knowledge integration is a fraught issue, involving conflict over knowledge, interests, and values (Diver, 2017; Boehensky & Maru, 2011), often creating binaries where Indigenous knowledge and science are pitted against each other (Evering, 2012). Traditional approaches to integration are also critiqued for ignoring the role of power relations and for prioritising certain kinds of knowledge over others, leading to knowledge products that serve science and the state over the interests of Indigenous communities. Attempting to 'integrate' both paradigms can create a management tension, as in practice each knowledge system operates in entirely different ways (Nursey-Bray & Arabana Corporation, 2015).

'Integration' also implies the adopting or incorporation of Indigenous knowledge into current jurisdictional and institutional arrangements – but they simply may not fit. This is partly due to the reality that Indigenous conceptions of knowledge, climate and adaptation cannot be compartmentalised into the various dimensions of 'management' aligned with common law that often occurs today.

The key to respecting different knowledge domains is accepting the holistic connections that designate cultural affiliation to, responsibilities for, and knowledge and use about an area of land and sea. A connection, that in many Indigenous contexts reflects an indivisibility between people and place: Indigenous peoples are themselves connected to it, are part of, and thus seamlessly connected to their regions. As such, the splitting and distinction between different parts of that region into 'policy' arenas, does not align with traditional modes of governance, nor knowledge keeping and dissemination in managing those lands and seas.

There is thus, no need to always 'integrate' but to seek ways in which Indigenous and non-Indigenous peoples can build bridges between, and to, each other. We suggest that knowledge *co-production* is one useful response in this context. It is one that builds knowledge in ways that transgress disciplinary boundaries and includes multiple knowledge realms. It is jointly produced and agreed to, the result of partnerships between societal actors, such as scientists and Indigenous peoples. Co-produced knowledge at the knowledge interface challenges entrenched norms of knowledge, including the sciences and increases its legitimacy, relevance and usability (Jasanoff, 2004). Co-production of knowledge via a partnership between Ltyente Apurte Rangers and staff from the Central Land Council in Australia has resulted in the co-creation of climate change presentations in the local Arrernte language and also the identification of potential adaptation actions (Hill et al., 2020). Successful co-production approaches such as these, yield insights into how context affects outputs and outcomes, identify how drivers and barriers to success differ, and challenge the hegemony of particular ways of knowing.

However, ultimately, we suggest that the notion of co-existence is perhaps a better 'fit' and provides a way in which to conceive how multiple knowledge may work together: sharing space and place in just, sustainable and equitable ways (Coombes et al., 2013). Knowledge *co-existence* can build on these insights to provide a pathway by which different forms of knowledge can co-exist at the knowledge interface (Nursey-Bray & Arabana Aboriginal Corporation, 2015). As Howitt (2020) argues knowledge co-existence is a means of acknowledging colonial invasion and the enabling practices of mutual recognition, collaborative building of consent, and appreciation of cultural continuity. Indigenous scholar Whyte (2020) additionally notes that such relational qualities are crucial for cross-societal coordination: and builds responsibility and collective action (Whyte, 2014, 2017). Trust, strong standards of consent, and genuine expectations of reciprocity can, in concert, act to build conservation outcomes.

Processes of knowledge co-existence also provide the opportunity to create new inter-cultural arrangements that both acknowledge different knowledge systems while finding common ground (Rea & Messner, 2008, 86). As such, Indigenous interests and values about adaptation, can co-exist *alongside*, rather than being integrated *within*, other rights and interests. This approach implies a recognition that both parties are equal and that all knowledge systems are legitimate and valid. It is a process that does not focus on seeking 'the' answers but on how to trigger engagement between multiple knowledge systems, to create modes of co-existence in adaptation management that harnesses the essential dynamism within and between each system (Byg & Salick, 2009).

Indigenous peoples have been managing complex problems within their territories for much longer than western society and they still hold valuable knowledge for dealing with complex societal problems (Pert et al., 2015). Innovative governance changes will enable processes of co-production of adaptation systems that can bring together multiple parties to co-jointly adapt to climate change, build wellbeing, while recognising difference (Jordan et al., 2010). Allowing spaces for the development of what Artelle et al. (2019) refer to as the resurgence of Indigenous led governance in environmental management, enables innovative and socially just ways forward.

Agency and Survival

Ultimately, our case studies and collaborations have revealed, notwithstanding huge challenges, a compelling and positive story of Indigenous agency and survival. Agency is a challenging term and can be qualified in a range of ways: Ahearn (2001) provides a working definition of agency as: "the socio-culturally mediated capacity to act". Agency can also be constructed as resistance to power (Frank, 2006), the "flexible wielding of means towards ends" (Kockelman, 2007); where power and knowledge intersect in dynamic ways. Agency is the mediating factor between them, and manifest as the push and pull between the individual and structure of the state. As such, agency might initially be understood as the relatively flexible wielding of means toward ends, and as such can also be seen as a form of resistance.

Of relevance to the Indigenous context, Dissanayake (1996, ix) states that agency is also about understanding the "historical and cultural conditions that facilitate the discursive production of agency, and on useful ways of framing the question of agency" and that doing so would enable better understanding of the "contours of the cultures that we study". In this context, resistance is the product of the interplay between multiple subject-positions. Cairns (2009) builds on this idea, noting that structure/agency can be differentiated where one is about individual, collective and immediate action, while the other is systemic, anonymised and bureaucratic. As such, agency is conceptualised as the tension between the individual and the social, political and economic structures that can constrain those (Cairns, 2009). Indigenous political agency can be based on multiple forms of power, changeable over time, embody multiple sites of encounters and powers, and can be produced across and within multiple agencies (Tennberg, 2010).

Agency is also asserted through Indigenous innovation, where old ways really do provoke new days: via the transformation of traditional knowledge into future adaptations.

Ultimately, Indigenous groups situate current climate projections within a discourse around an age-old trajectory of change and adaptation; where livelihoods equalled survival and adaptations have been experienced for millennia. Hence, climate change, and twenty first century projections, are situated within a temporal continuum that reflects enduring survival to multiple climatic changes. Adaptation is not just adapting to climate change but ALL change and is embodied within Indigenous insistence on documenting all changes, not just climatic ones. Contemporary adaptation partnerships would benefit from the integrated perspective this insight brings.

A Last Reflection

Indigenous peoples across the world play a key role in global conservation and increasingly in addressing the impacts of climate change. Globally, 370 million people identify as Indigenous, and manage or have tenure rights over 38 million square kilometres (Garnett et al., 2018). This represents over a quarter of the world's land surface and includes multiple protected and other environmentally significant areas (Garnett et al., 2018). Further, Indigenous peoples now also live in multiple cities and regions.

Overall, we find that despite the incredible hardships faced by Indigenous groups across the world there is a current and growing wealth of Indigenous led adaptation that reflects agency and will to survive in the midst of the climate crisis. However, adaptation is not simply a reply to the call to action on climate change, it also, in many cases, such as in Australia and the United States, represents a palimpsest of Indigenous voices that seek to deploy adaptation as a vehicle for the redressing of the hurt and destruction caused by colonisation, and today globalisation. Their collective knowledge and histories have provided a basis from which to develop culturally driven adaptation that will ensure their survival.

We have demonstrated that Western climate terminology can be used to entrench existing inequalities, and to prioritise Western knowledge over other knowledge, yet we also show how, in their rejection of these terms, and subsequent articulation of their own discourses about change, Indigenous peoples assert power and agency, as well as their capacity to survive and change. We do not present idealised ideas about what Indigeneity is, nor will all of the topics and issues raised in this book apply to all peoples, but we suggest that rather than pre-judging how and what Indigenous adaptation is, that we learn to listen, hear, observe and empower every Indigenous group to decide what it is for themselves.

We conclude that Indigenous adaptation is not just about developing a program that employs one or two Indigenous people. Nor is it about documenting their knowledge, or supporting them to develop a plan, that is then not supported in practice or implementation because it does not 'fit'. Nor is it about responding to the friendly overtures of policy brokers who seek to work with Indigenous peoples to document – and then potentially appropriate – their knowledge for implementation into Western adaptation programs. It is about developing co-existence: the co-production of adaptation systems, ones that promote inter-sectoral and context specific coordination in ways that can enjoin many parties to co-jointly manage the impacts of and build adaptive solutions to climate change. It is about embracing all forms of knowledge not just the ones deemed culturally palatable by non-Indigenous voices.

Ultimately it is about creating spaces and places for Indigenous peoples at the decision-making tables, and about supporting Indigenous driven and crafted governance and policy responses. These spaces will enhance Indigenous wellbeing by ensuring equality while recognising cultural difference. Such a collaborative decision-making system would also need to recognise different Indigenous knowledge systems (pre-and postcolonial), in ways that encourage gender equity and facilitate hybrid economies to build community capacity and resilience. This is an area that is ripe for future research: the development of modes of co-existence that incorporate cultural differences, histories and that will facilitate the opening of spaces in policy regimes that support and advance Indigenous voices at all levels of decision making.

Indigenous people face enormous challenges in navigating climate change. They cannot do it alone, yet as the examples and case studies in this book have shown, they are not passive victims of global environmental change and are actively seeking to adapt or cope with the changing climatic conditions. Indigenous adaptation is both a *process*, enacted over many ages, and a *practice*, where in the here and now, old ways have been adapted to new ways, to build contemporary adaptation programs and new futures (Photo 7.2).

Photo 7.2 Sturt Desert Pea, Arabana Country, Australia. (Credit: Melissa Nursey-Bray)

References

Ahearn, L. M. (2001). Language and agency. *Annual Review of Anthropology, 30*(1), 109–137.

Artelle, K. A., Zurba, M., Bhattacharyya, J., Chan, D. E., Brown, K., Housty, J., & Moola, F. (2019). Supporting resurgent indigenous-led governance: A nascent mechanism for just and effective conservation. *Biological Conservation, 240*, 108284.

Barber, M. (2018). Indigenous knowledge, local history and environmental change amongst the Yolngu people of Blue Mud Bay, Arnhem Land, Australia. In D. Nakashima (Ed.), *Indigenous knowledge for climate change assessment and adaptation* (pp. 16–122). UNESCO-Cambridge University Press.

Barber, M. (2010). Coastal conflicts and reciprocal relations: Encounters between Yolngu people and commercial fishermen in Blue Mud Bay, north-east Arnhem Land, *The Australian Journal of Anthropology, 21*(3), 298–314.

Bardsley, D. K. (2018). Indigenous knowledge and practice for climate change adaptation. In D. A. Della Sala & M. I. Goldstein (Eds.), *Encyclopedia of the anthropocene* (Vol. 2, pp. 359–367). Elsevier.

Bardsley, D. K., & Wiseman, N. D. (2012). Climate change vulnerability and social development for remote indigenous communities of South Australia. *Global Environmental Change, 22*(3), 713–723.

Bardsley, D. K., Moskwa, E., Weber, D., Robinson, G. M., Waschl, N., & Bardsley, A. M. (2018). Climate change, bushfire risk, and environmental values: Examining a potential risk perception threshold in peri-urban South Australia. *Society & Natural Resources, 31*(4), 424–441.

Berkes, F. (2009). Indigenous ways of knowing and the study of environmental change. *Journal of the Royal Society of New Zealand, 39*(4), 151–156.

Berkes, F., Colding, J., & Folke, C. (2000). Rediscovery of traditional ecological knowledge as adaptive management. *Ecological Applications, 10*(5), 1251–1262.

Bohensky, M., & Maru. (2011). Indigenous knowledge, science, and resilience: What have we learned from a decade of international literature on "integration"? *Ecology and Society, 16*(4), 6.

Byg, A., & Salick, J. (2009). Local perspectives on a global phenomenon—Climate change in eastern Tibetan villages. *Global Environmental Change, 19*(2), 156–166.

Cairns, S. (2009). Agency. *Architectural Research Quarterly, 13*(2), 105–108.

Cameron, E. S. (2012). Securing Indigenous politics: A critique of the vulnerability and adaptation approach to the human dimensions of climate change in the Canadian Arctic. *Global Environmental Change, 22*(1), 103–114.

Cockerill, J. (2018, 22 March). *Hauntology on country*. Overland.

Conedera, M., Corti, G., Piccini, P., Ryser, D., Guerini, F., & Ceschi, I. (2004). La gestione degli incendi boschivi in Canton Ticino: tentativo di una sintesi storica/Forest fire management in Canton Ticino: attempting a historical overview. *Schweizerische Zeitschrift fur Forstwesen, 155*(7), 263–277.

Conedera, M., Vassere, S., Neff, C., Meurer, M., & Krebs, P. (2007). Using toponymy to reconstruct past land use: A case study of 'brüsáda'(burn) in southern Switzerland. *Journal of Historical Geography, 33*(4), 729–748.

Conedera, M., Colombaroli, D., Tinner, W., Krebs, P., & Whitlock, C. (2017). Insights about past forest dynamics as a tool for present and future forest management in Switzerland. *Forest Ecology and Management, 388*, 100–112.

Coombes, B., Johnson, J. T., & Howitt, R. (2013). Indigenous geographies II: The aspirational spaces in postcolonial politics – Reconciliation, belonging and social provision. *Progress in Human Geography, 37*(5), 691–700.

Dissanayake, W. (Ed.). (1996). *Narratives of agency: Self-making in China, India and Japan*. University of Minnesota press.

Diver, S. (2017). Negotiating Indigenous knowledge at the science-policy interface: Insights from the Xáxli'p Community Forest. *Environmental Science & Policy, 73*(C), 1–11.

Etchart, L. (2017). The role of indigenous peoples in combating climate change. *Palgrave Communications, 3*(1), 1–4.

Evering, B. (2012). Relationships between knowledge(s): Implications for "knowledge integration". *Journal of Environmental Studies and Sciences, 2*(4), 357–368.

Fernández-Llamazares, Á., Garcia, R. A., Díaz-Reviriego, I., Cabeza, M., Pyhälä, A., & Reyes-García, V. (2017). An empirically tested overlap between indigenous and scientific knowledge of a changing climate in Bolivian Amazonia. *Regional Environmental Change, 17*(6), 1673–1685.

Fletcher, S. M., Thiessen, J., Gero, A., Rumsey, M., Kuruppu, N., & Willetts, J. (2013). Traditional coping strategies and disaster response: Examples from the South Pacific region. *Journal of Environmental and Public Health, 2013*, 1–9.

Frank, K. (2006). Agency. *Anthropological Theory, 6*(3), 281–302.

Garnett, S. T., Burgess, N. D., Fa, J. E., Fernández-Llamazares, Á., Molnár, Z., Robinson, C. J., Watson, J., Zander, K., Beau, A., Eduardo, S., Collier, N., French, N., Duncan, T., Ellis, E., Geyle, H., Jackson, M., Jonas, H., Malmer, P., McGowan, B., … Leiper, I. (2018). A spatial overview of the global importance of indigenous lands for conservation. *Nature Sustainability, 1*(7), 369–374.

Gyampoh, B. A., Amisah, S., Idinoba, M., & Nkem, J. (2009). Using traditional knowledge to cope with climate change in rural Ghana. *Unasylva, 60*(281/232), 70–74.

Hill, R., Çiğdem, A., Wilfred, V. A., Zsolt, M., Yildiz, A., Bridgewater, P., Tengö, M., Thaman, R., Adou Yao, C., Berkes, F., Carino, J., Carneiro da Cunha, M., Diaw, M. C., Díaz, S., Figueroa, V., Fisher, J., Hardison, P., Ichikawa, K., Kariuki, P., … Xue, D. (2020). Working with Indigenous, local and scientific knowledge in assessments of nature and nature's linkages with people. *Current Opinion in Environmental Sustainability, 43*, 8–20.

Howitt, R. (2020). Unsettling the taken (for granted). *Progress in Human Geography, 44*(2), 193–215.

Howitt, R., & Klaus, F. (2012). Sustainable indigenous futures in remote Indigenous areas: Relationships, processes and failed state approaches. *GeoJournal, 77*(6), 817–828.

Hunt, J., Smith, D., Garling, S., & Sanders, W. (2008). *Contested governance. Culture, power and institutions in indigenous Australia*. CAEPR, ANU, Australia.

IPCC. (2014). *Annex II: Glossary in: Climate change 2014: Impacts, adaptation, and vulnerability. Contribution of working group II to the fifth assessment report of the Intergovernmental Panel on Climate Change* (pp. 1757–1776). Cambridge University Press.

Jasanoff, S. (2004). *States of knowledge: The co-production of science and the social order, states of knowledge*. Routledge.

Johnson, J. T., & Murton, B. (2007). Re/placing native science: Indigenous voices in contemporary constructions of nature. *Geographical Research, 45*(2), 121–129.

Jordan, K., Bulloch, H., & Buchanan, G. (2010). Statistical equality and cultural difference in Indigenous wellbeing frameworks: A new expression of an enduring debate. *Australian Journal of Social Issues, 45*(3), 333–362.

Kockelman, P. (2007). Agency: The relation between meaning, power, and knowledge. *Current Anthropology, 48*(3), 375–401.

Kuruppu, N., & Liverman, D. (2011). Mental preparation for climate adaptation: The role of cognition and culture in enhancing adaptive capacity of water management in Kiribati. *Global Environmental Change, 21*(2), 657–669.

Mathiesen, S., Bongo, M., Burgess, P., Corell, R., Degteva, A., Eira, I., & Vikhamar-Schuler, D. (2018). Indigenous reindeer herding and adaptation to new hazards in the Arctic. In D. Nakashima, I. Krupnik, & J. T. Rubis (Eds.), *Indigenous knowledge for climate change assessment and adaptation* (pp. 198–213). Cambridge University Press.

Matuk, F. A., Behagel, J. H., Simas, F. N. B., Do Amaral, E. F., Haverroth, M., & Turnhout, E. (2020). Including diverse knowledges and worldviews in environmental assessment and planning: The Brazilian Amazon Kaxinawá Nova Olinda indigenous land case. *Ecosystems and People (Abingdon, England), 16*(1), 95–113.

Mistry, J., Bilbao, B. A., & Berardi, A. (2016). Community owned solutions for fire management in tropical ecosystems: Case studies from Indigenous communities of South America. *Philosophical Transactions of the Royal Society B: Biological Sciences, 371*, 1–10.

Nakashima, D., Krupnik, I., & Rubis, J. T. (2018). *Indigenous knowledge for climate change assessment and adaptation*. Cambridge University Press.

Nkomwa, E. C., Joshua, M. K., Ngongondo, C., Monjerezi, M., & Chipungu, F. (2014). Assessing indigenous knowledge systems and climate change adaptation strategies in agriculture: A case study of Chagaka Village, Chikhwawa, Southern Malawi. *Physics and Chemistry of the Earth. Parts A/B/C, 67–69*, 164–172.

Nursey-Bray, M., & Arabana Aboriginal Corporation. (2015). Cultural indicators, country and culture: The Arabana, change and water. *The Rangeland Journal, 37*(6), 555.

Nursey-Bray, M., & Palmer, R. (2018). Country, climate change adaptation and colonisation: Insights from an Indigenous adaptation planning process, Australia. *Heliyon, 4*(3), e00565.

Nursey-Bray, M., Fergie, D., Arbon, V., Rigney, L. I., Palmer, R., Tibby, J., Harvey, N., Hackworth, L., & Stuart, A. (2015). Indigenous adaptation to climate change: The Arabana. In J. Palutikof, S. Boulter, J. Barnett, & D. Rissik (Eds.), *Applied studies in climate adaptation* (pp. 316–325). Wiley, West Sussex. https://doi.org/10.1002/9781118845028.ch35

Nursey-Bray, M., Palmer, R., Stuart, A., Arbon, V., & Rigney, L. E. (2020). Scale, colonisation and adapting to climate change: Insights from the Arabana people, South Australia. *Geoforum, 114*, 138–150.

Nursey-Bray, M., Lui, M., Lui, M., Malsale, P., Mariner, A., Nelson, F., Nihmei, S., Parsons, M., Ronneberg, E., & Jerome, S. (2021). Adapting to change: Traditional knowledge and water. In B. Dansie, H. Böer, & H. Alleway (Eds.), *Securing water, energy and food in the Pacific*. UNESCO.

Pert, P. L., Ens, E. J., Locke, J., Clarke, P. A., Packer, J. M., & Turpin, G. (2015). An online spatial database of Australian indigenous biocultural knowledge for contemporary natural and cultural resource management. *The Science of the Total Environment, 534*, 110–121.

Porter, L., & Davoudi, S. (2012). The politics of resilience for planning: A cautionary note: Special issue edited by S. Davoudi and L. Porter (eds). *Planning Theory & Practice, 13*(2), 299–333.

Race, D., Mathew, S., Campbell, M., & Hampton, K. (2016). Understanding climate adaptation investments for communities living in desert Australia: Experiences of indigenous communities. *Climatic Change, 139*(3–4), 461–475.

Rea, N., & Messner, C. (2008). *The character of Aboriginal Training Pathways: A local perspective*, Desert Knowledge Cooperative Research centre, Alice Springs, Australia, vi, 29.p.

Reo, N., Topkok, S. M., Kanayurak, N., Stanford, J., Peterson, D., Whaley, L., & Giguère, N. (2019). Environmental change and sustainability of indigenous languages in Northern Alaska. *Arctic, 72*(3), 215–228.

Robinson, D. F., & Raven, M. (2020). Recognising Indigenous customary law of totemic plant species: Challenges and pathways. *The Geographical Journal, 186*(1), 31–44.

Roos, C. I., Zedeño, M. N., Hollenback, K. L., & Erlick, M. M. (2018). Indigenous impacts on North American Great Plains fire regimes of the past millennium. *Proceedings of the National Academy of Sciences, 115*(32), 8143–8148.

Sánchez-Cortés, M. S., & Chavero, E. L. (2011). Indigenous perception of changes in climate variability and its relationship with agriculture in a Zoque community of Chiapas, Mexico. *Climatic Change, 107*(3–4), 363–389.

Schreiber, D. (2006). First nations, consultation, and the rule of law: Salmon farming and colonialism in British Columbia. *American Indian Culture and Research Journal, 30*(4), 19–40.

Tebrakunna country, & Lee, E. (2019). 'Reset the relationship': Decolonising government to increase indigenous benefit. *Cultural Geographies, 26*(4), 415–434.

Tengö, M., Brondizio, E. S., Elmqvist, T., Malmer, P., & Spierenburg, M. (2014). Connecting diverse knowledge Systems for Enhanced Ecosystem Governance: The multiple evidence base approach. *Ambio, 43*(5), 579–591.

Tennberg, M. (2010). Indigenous peoples as international political actors: A summary. *The Polar Record, 46*(3), 264.

Tinner, W., Hubschmid, P., Wehrli, M., Ammann, B., & Conedera, M. (1999). Long-term forest fire ecology and dynamics in southern Switzerland. *Journal of Ecology, 87*(2), 273–289.

Tinner, W., Conedera, M., Ammann, B., & Lotter, A. F. (2005). Fire ecology north and south of the Alps since the last ice age. *The Holocene, 15*(8), 1214–1226.

Veland, S., Howitt, R., Dominey-Howes, D., Thomalla, F., & Houston, D. (2013). Procedural vulnerability: Understanding environmental change in a remote indigenous community. *Global Environmental Change, 23*(1), 314–326.

Weichselgartner, J., & Kelman, I. (2015). Geographies of resilience: Challenges and opportunities of a descriptive concept. *Progress in Human Geography, 39*(3), 249–267.

Whyte, K. P. (2014). Indigenous women, climate change impacts, and collective action. *Hypatia, 29*(3), 599–616.

Whyte, K. (2017). Indigenous climate change studies: Indigenizing futures, decolonizing the Anthropocene. *English Language Notes, 55*(1–2), 153–162.

Whyte, K. (2020). Too late for indigenous climate justice: Ecological and relational tipping points. *Wiley Interdisciplinary Reviews. Climate Change, 11*(1), 1–7.

Wildcat, D. R. (2013). Introduction: Climate change and indigenous peoples of the USA. *Climatic Change, 120*(3), 509–515.

Wiseman, N., & Bardsley, D. (2016). Monitoring to learn, learning to monitor: A critical analysis of opportunities for indigenous community-based monitoring of environmental change in Australian rangelands. *Geographical Research, 54*(910), 52–71.

Index

© The Author(s) 2022
M. Nursey-Bray et al., *Old Ways for New Days*, SpringerBriefs in Climate Studies, https://doi.org/10.1007/978-3-030-97826-6

CPSIA information can be obtained
at www.ICGtesting.com
Printed in the USA
LVHW081212060922
727579LV00012B/31